P9-AFH-704

Radiation Risks in Perspective

Kenneth L. Mossman

Taylor & Francis
Taylor & Francis Group
Boca Raton London New York

CRC is an imprint of the Taylor & Francis Group,
an informa business

CRC Press
Taylor & Francis Group
6000 Broken Sound Parkway NW, Suite 300
Boca Raton, FL 33487-2742

Printed in the United States of America on acid-free paper
10 9 8 7 6 5 4 3 2 1

International Standard Book Number-10: 0-8493-7977-6 (Hardcover)
International Standard Book Number-13: 978-0-8493-7977-2 (Hardcover)

Library of Congress Cataloging-in-Publication Data

Mossman, Kenneth L., 1946-
Radiation risks in perspective / by Kenneth L. Mossman.
p. cm.
Includes bibliographical references (p.).
ISBN 0-8493-7977-6 (alk. paper)
1. Radiation carcinogenesis. 2. Health risk assessment. I. Title.

RC268.55.M673 2006
616.99'4071--dc22 2006047553

Visit the Taylor & Francis Web site at
http://www.taylorandfrancis.com

and the CRC Press Web site at
http://www.crcpress.com

Dedication

To my wife, Blaire —
Who has always been there for me and supported me in ways that cannot be expressed in words
To my parents, Meyer David Mossman (1915–1995) and Sarah Kutchai Mossman (1920–2005) —
May your memory be for a blessing

Table of Contents

Author xi

Acknowledgments xiii

Introduction xv

Abbreviations xxv

List of Figures xxvii

List of Tables xxix

Chapter 1 Risky Business 1

More than a Number 3
Safety without Risk? 6
What's Risky? 7
Is It Dangerous? 9
Can I Get Exposed? 13
Can It Hurt Me? 15
What Are the Risks? 16
Damage Control 17
Perception Is Reality 19
Notes and References 21

Chapter 2 Scientific Guesswork 25

Making the Right Choice 26
Predictive Theories in Risk Assessment 30
    Linear No-Threshold Theory 30
    Sublinear Nonthreshold 31
    Supralinear 32
    Hormesis 32
    Threshold 34
Limitations and Uncertainties 36
Speculation versus Reality 38
Risk Management and Risk Communication 39
Quantifying Risk at Small Doses 41
Notes and References 42

Chapter 3 No Safe Dose 47

LNT: The Theory of Choice 48
The LNT Controversy 53
Elements of the Debate 54
The Question of Thresholds 54
Repair of Radiation Damage and Cellular Autonomy 54
Uses and Misuses of LNT 55
Case 1: Estimation of Health Effects of Fallout from the Chernobyl Reactor Accident 56
Case 2: Childhood Cancer Following Diagnostic X-Ray 56
Case 3: Public Health Impacts from Radiation in a Modern Pit Facility 58
LNT Consequences 58
Notes and References 60

Chapter 4 Uncertain Risk 65

How Low Can You Go? 66
Risk Assessment Considering Uncertainty 70
Uncertain Choices 72
Another Approach 74
Notes and References 75

Chapter 5 Zero or Bust 79

Management Triggers 81
Technical Triggers 82
Size Matters 83
Sensitive People 84
Assigned Blame 86
Social Triggers 87
Safety 88
Protection of Children and the Unborn 89
Polluters Should Pay 90
Catastrophe and Apathy 90
Public Information and Distorting Risks 90
Political Triggers 91
Perceptions and Conflicts of Interest 92
Management Strategies 94
As Low As Reasonably Achievable (ALARA) 94
Best Available Technology (BAT) 95
The Precautionary Principle 95
Risk–Risk Trade-offs and Unintended Consequences 99

Risk Offset 100
Risk Substitution 100
Risk Transfer and Risk Transformation 101
Challenges 101
Notes and References 103

Chapter 6 Misplaced Priorities 109

Priorities and Realities 110
Factors in Prioritization 112
Scientific Evidence 113
Public Perception of Risks 114
Management Capacity 117
Court Actions 118
Influence of Stakeholder Groups 119
Real Risks and Reordering Priorities 121
Monetary Costs 123
Environmental Cleanup at the Nevada Test Site 124
Characterization of Waste Destined for WIPP 125
Risks in Perspective 125
Notes and References 126

Chapter 7 Avoiding Risk 129

The Case against Risk 130
Different Risks 130
Agent–Agent Interactions 131
Dose as a Surrogate for Risk 132
The Case for Dose 133
A Dose-Based System of Protection 136
Regulatory Dose Limit 136
Natural Background 137
Acceptable Dose 139
Management Decisions Based on Dose Proportion 140
Simplification of Radiation Quantities and Units 142
Review of the Current System of Radiation Protection 144
Notes and References 145

Chapter 8 Radiation from the Gods 149

The Watras Case 150
Human Exposure to Radon 152
Health Hazards of Radon 153

Is There Really a Public Health Hazard? 156
Public Health 156
Perceptions and Fears 158
Economic Impacts 160
National/Regional Differences 162
Smoking Is the Problem 163
Notes and References 164

Chapter 9 Hold the Phone 167

Will Cell Phones "Fry" Your Brain? 168
Managing Phantom Risks 172
Imprudent Precaution 173
International Calls 177
Notes and References 179

Chapter 10 PR Campaign: Proportion, Prioritization, and Precaution 183

Proportion 184
Prioritization 187
Precaution 189
Notes and References 191

Glossary 193

Index 197

Author

Kenneth L. Mossman is a professor of health physics in the School of Life Sciences and affiliated faculty member of the Center for the Study of Law, Science and Technology at Arizona State University in Tempe, where he has also served as assistant vice president for research and director of the university's Office of Radiation Safety. Prior to his arrival at Arizona State University, Dr. Mossman was a faculty member of the medical and dental schools at Georgetown University in Washington, DC, and was professor and founding chairman of the Department of Radiation Science at Georgetown's Graduate School. His research interests include radiological health and safety and public policy. Dr. Mossman has authored more than 150 publications related to radiation health issues. He served as president of the Health Physics Society and received its prestigious Elda Anderson Award, the Marie Curie Gold Medal, and the Founder's Award. He has been a Sigma Xi distinguished lecturer and is a fellow of the Health Physics Society and the American Association for the Advancement of Science. He has served on committees of the National Research Council, the National Institutes of Health, the U.S. Nuclear Regulatory Commission, the Nuclear Energy Agency of the Organization for Economic Cooperation and Development (Paris), and the International Atomic Energy Agency (Vienna). Dr. Mossman earned a BS in biology from Wayne State University, MS and PhD degrees in radiation biology from the University of Tennessee, and an MEd degree in higher education administration from the University of Maryland. Dr. Mossman is also author of *The Radiobiological Basis of Radiation Protection Practice* (1992) Lippincott Williams & Wilkins, with William Mills and *Arbitrary and Capricious* (2004) AEI Press, Washington, DC with Gary Marchant.

Acknowledgments

Many people contributed to this book by stimulating ideas or helping to clarify my own thinking. Ideas are never created in a vacuum but, instead, are products of work that went on before. Creative ideas come from unique configurations and insights into past ideas and observations in the context of current times. A good idea now may not necessarily be a good idea in the future or a good idea for the past. I have relied heavily on the work of others in developing the themes and ideas for this book. The detailed references and notes at the end of each chapter reflect this.

I am particularly grateful to the following colleagues and former students who provided stimulating discussions and sharpened my thinking about science, policy, law, radiation protection, and radiobiology: Allen Brodsky, Antone Brooks, Bernard Cohen, Keith Dinger, John Frazier, Raymond Guilmette, C. Rick Jones, Cynthia G. Jones, Edward Lazo, Sigurdur Magnusson, Gary Marchant, Henri Metivier, Kenneth Miller, William Mills, Alan Pasternak, Otto Raabe, Keith Schiager, Chauncey Starr, Richard Vetter, and Chris Whipple.

I am particularly indebted to Michael Ryan and Tim Jorgensen, who critically reviewed an early draft of the manuscript and provided excellent suggestions. Their efforts improved the work immeasurably.

I take full responsibility for statements made in this book. I took great care to properly credit and accurately reflect the views and findings of others. Any errors, omissions, or misrepresentations are my fault and entirely unintentional.

I am grateful to the John Simon Guggenheim Memorial Foundation for their generous fellowship to support background research and early writing of this book.

I thank my editor, Cindy Carelli, and project coordinator, Jill Jurgensen at Taylor & Francis, for their support of the book; for their assistance in the final editing stages of the manuscript; and for ably orchestrating book production. I am particularly grateful to Cindy for believing in this book and shepherding my proposal through the review process.

Finally, I thank my wife Blaire, who has been a pillar of strength throughout the manuscript-writing process. She served as a sounding board for many ideas, and her perspectives were extremely valuable to me in writing several chapters. As a professional editor, Blaire's expertise was invaluable in organizing the book. Blaire read the entire manuscript and her technical editing skills clarified and markedly improved my writing.

Introduction

How people deal with risks is one of the most interesting aspects of the human condition. Perceptions of risk are complex, and people deal with risks in different, often irrational, ways. Young, healthy people consider themselves immortal and immune from long-term, degenerative diseases. As a consequence they engage in risky dietary and smoking behaviors that enhance the chances of such diseases in later life.

People often react irrationally to safety measures and to environmental health threats. Some drivers and passengers continue to avoid using seat belts even though it is well known that car restraints are one of the cheapest and most effective ways to reduce injuries and fatalities.

Risk is difficult to comprehend, particularly when probabilities of occurrence are very small. A one in a million risk is beyond common understanding because such probabilities are outside everyday experiences. Not many people have direct experience with a one in a million risk. The public cannot distinguish real from phantom events that may occur with very low frequency. Many people play the lottery in the hope of winning a multimillion-dollar jackpot that has astronomical odds of winning. Lottery participation is fueled by the fact that someone has to win. Unfortunately, the same thinking goes into public perception of cancer risks — someone has to get cancer even when the risks are very small. Very low risks of cancer (say one in a million) associated with consuming pesticide residues in foods are considered real because of the belief that if enough people are exposed someone will get cancer from pesticides. Unlike lottery probabilities, very small cancer risks are determined theoretically. In lotteries someone eventually wins, but very small cancer risks do not mean someone will get cancer. The public overestimates the magnitude of the risks. Perception and scientific understanding are incongruent. The public cannot distinguish real risks from theoretical ones.

The consequences are serious. For many technological risks, the public's view is out of touch with what is actually known about the probability and consequences of the risk. The public is wary of advanced, sophisticated technologies and their risks because they are not understood very well. Furthermore, benefits of the technologies may not be obvious. The public has little or no tolerance for environmental contamination and health risks because of the perception of lack of public control or corporate responsibility. Obviously, risk perception depends on many psychosocial factors; some groups may find otherwise intolerable risks acceptable because benefits are clear (e.g., employment, strong local economy). In general, technological risks, no matter how small, are viewed as unacceptable and must be strictly regulated. If a company polluted the environment, then it should be responsible for cleaning up its mess. But the costs of reducing already small risks can be enormous, and the environmental and public health benefits may be questionable. As the public's tolerance for risk diminishes, regulatory compliance costs grow exponentially, and it's the public that ends up paying the bill through increased costs of goods and services.

This book focuses on small risks associated with environmental and occupational exposures to physical and chemical carcinogens. Cancer is an important health outcome. It is the second leading cause of death in the U.S. Cancer is the major long-term health effect of concern to government agencies charged with protecting worker and public health. Emphasis in this book is on ionizing radiation because more is known about its cancer-causing effects than almost any other agent. Unlike many chemical agents, there is substantial experience with human exposure over a wide range of doses. Concepts and principles in risk assessment and risk management were derived in large part from experiences with ionizing radiation. Nuclear and radiological technologies, including nuclear power plants and medical imaging, are two of the most tightly regulated industries, and they have been a consistent source of public controversy. Issues such as health and environmental effects of radioactive waste repositories and cancer risks from routine medical x-rays remain high on the public's radar screen.

For the purposes of discussion, small risk is defined as a probability of 1 in 10,000 or less dying of cancer from exposure to a carcinogenic agent. Such risks cannot be observed directly because the spontaneous incidence of cancer is very high and the probability of cancer from agent exposure is very small by comparison. Therefore, risk estimates must be derived theoretically by extrapolating from cancer induction at high doses of the agent. The public tends to overestimate small cancer risks because of fear of cancer and difficulties comprehending small probabilities.

Radiation Risks in Perspective calls for rethinking how small risks are measured, communicated, and managed. Do small increments in doses above what occurs naturally result in discernible increases in risks? Given that the public has difficulty understanding small probabilities, is the concept of "risk" the most appropriate way to express potential health detriment? Regardless of the metric used, small risks should not be ignored. If small risks can be managed, we should do so while recognizing that economic resources are limited. The public needs to reexamine risk priorities and focus on the major risks that are responsible for public health problems such as smoking, diet, and physical inactivity. These factors are recognized as the major contributors to the three major causes of mortality in the U.S. — heart disease, cancer, and diabetes. Control of large risks should trump concerns over minor risks. Although we may have the capacity to manage very small risks, this does not mean we should focus disproportionate resources toward solving small problems.

How much attention should we pay to small risks? Clearly this depends on the nature of the risks, what caused the exposure, and how readily the exposure and consequences can be managed. Risks with small probabilities of occurrence can be serious if the consequences are severe and large numbers of people are affected. The Sumatra tsunami of December 2004 is an example of a small probability event that had devastating consequences. But phantom risks can also generate public health crises. The 1989 apple scare is an example of how phantom risks can be used to manufacture a nonexistent public health crisis and effectively shut down an important food industry for months. The Natural Resources Defense Council (NRDC) spearheaded an effort to ban the use of Alar because it might cause cancer. Based on very limited animal data and no human experience, the risk of cancer in humans would be about 1 in 100 million if a person ate one apple a day for 70 years. This risk is so small that it is essentially nonexistent. But the U.S. Environmental Protection Agency (EPA) banned Alar anyway and the apple-growing industry lost more than $200 million. The Alar case also illustrates the fallacy of summing small risks over very large populations to arrive at a measure of public health detriment. No one will get cancer from apples sprayed with Alar although an individual risk of 1 in 100 million translates into 3 cancers in the U.S. population of 300 million.

Risks to the individual do not necessarily translate into a population risk. An individual with a one in a million risk of cancer due to exposure to a cancer-causing agent has that risk regardless of the population size even though one would predict, in statistical terms, 10 cases in a population of 10 million or 100 cases in a population of 100 million if everyone in the population received the same dose of the agent. It is entirely possible that no one in the population will get cancer from the agent. Such calculations suggest a public health problem simply because large populations are used in the calculation. If the individual isn't harmed by the agent, then the population isn't either. However, infectious diseases present an entirely different situation. An individual's chance of getting infected depends on the number of people already infected.

In calling for reforms to improve assessment and management of technological risks and public understanding of health risks, I explore three major themes in the book. In the first theme, I argue that risk should be replaced by dose as a basis for assessment and management decisions. The concept of dose proportion is introduced as a new metric for risk assessment and management. It is a dimensionless quantity that compares measured dose or exposure to an agent with an appropriate reference source. Working with dose instead of risk has decided advantages. Tiny amounts of radiation and chemicals can be reliably measured. Detection levels are orders of magnitude below levels required to observe health effects. The public readily comprehends the concept of dose based on everyday experiences with prescription and over-the-counter medications. Dose is not used as a surrogate for risk and therefore avoids the uncertainties inherent in quantifying risk. Management decisions are based on measured doses and comparisons with reference sources including natural background levels and regulatory dose limits.

A new paradigm must be developed to avoid use of predictive theories as a basis for quantifying small risks. The use of dose proportions obviates the need to use problematic predictive theories to estimate risks that are highly uncertain. Our knowledge of health risks from exposure to pesticides, pollutants, other chemicals, and ionizing radiation is based primarily on data derived from studies or experiences at high doses of the agent. In laboratory studies, large doses are used to increase the probability of observing the effect. Predictive theories such as the linear no-threshold theory (LNT) are used to estimate health risks at much smaller doses typically encountered in environmental or occupational settings. Under LNT theory, reducing the dose tenfold reduces the risk tenfold. However, theory-derived risks often have very large uncertainties that limit their value in decision making. What does a risk of 1 in 100,000 mean when the lower bound of uncertainty includes zero? Use of predictive theories to estimate small risks should also be avoided because of inherent uncertainties in the theories themselves. Furthermore, the public mistakenly concludes that theory-derived risks are "real" when, in fact, they are nothing more than "speculates."

There are other risk communication challenges. How risk is expressed by experts, technocrats, and the public is at the heart of the communication problem. The public does not understand very small probabilities. Scientists, engineers, and other "experts" who assess, manage, and analyze risks prefer to use a quantitative approach to describe risks. However, the public is generally ill prepared to deal with risk numbers, particularly if they are expressed as percentages or ratios that can be easily misinterpreted.

Risk is used as a "coin of the realm." The idea is that risk (calculated by converting doses to risks using a predictive theory like LNT) allows different agents to be compared and individual risks combined together to arrive at a single number that reflects total health detriment. The underlying assumption is that all carcinogens produce the same health outcome. As we have learned more about cancer and the factors that cause cancer, it has become quite apparent that using risk as a form of currency is seriously flawed. Cancer risks are not the same for all agents and cannot be compared or combined. Carcinogens such as smoking and ultraviolet radiation produce diseases that have different histories, clinical courses, and outcomes. The risk of skin cancer cannot be legitimately compared to the risk of lung cancer. Even when agents produce the same disease (e.g., smoking and ionizing radiation both increase the risk of lung cancer), combining risks is problematic because risks may not be strictly added if there is overlap in mechanisms of pathogenesis.

The second theme explores prioritization of risk. Which risks are important and which ones are insignificant? How are risks perceived and what does risk perception have to do with how we prioritize risks? Unfortunately, we fear the wrong things and spend our money to protect ourselves from the wrong dangers. There is a need to balance risk with benefits and with other risks. Society should focus on management of risks for which there may be substantial gain in benefit (e.g., reducing cigarette consumption). The idea that we need to rethink how we prioritize risks is not new. What is new in this book is the notion that analyzing and discussing individual risks without regard to the presence of other risks is inappropriate. Isolated risks appear to be important but may become less so when considered in the context of other risks in the environmental or occupational setting. Society should focus on risks reduction for which there may be substantial gain (e.g., cigarettes, radon reduction in homes with high indoor air concentrations). Of course, this is easier said than done. Given substantial limitations in resources, society is faced with some hard choices regarding the need to balance reduction in institutional risks (e.g., occupational exposures) against the larger public health challenges associated with individual risks, including diet, smoking, and physical inactivity.

The third theme concerns the use of precaution in risk management. The public continues to push toward zero risk tolerance in which a balanced cost-benefit approach to risk management is replaced by a precautionary approach of better safe than sorry. The precautionary principle states that when an activity or technology may harm human health or the environment, precautionary measures should be taken, including a ban or severe restrictions on the activity or technology, although risks may not be fully characterized. The precautionary principle is an extremist approach to risk management that can lead to unreasonable restrictions on technologies.

Precautionary approaches are reasonable for technologies in which activities or products are known to produce very serious risks with little benefit. The precautionary approach has an important role to play in cases such as global warming and proliferation of nuclear weapons where potential consequences are severe and eliminating certain technologies would diminish the risk. In radiological protection and in the management of risks from chemical carcinogens, there is no justification for implementation of the precautionary principle because properly controlled sources of radiation and chemicals do not pose significant threats to public health and the environment. Instead, risk management should focus on a balanced approach to costs and benefits. Technological risks must be controlled but not to the extent that social benefits are severely compromised. If regulatory compliance costs become excessive, goods and services are priced out of the market and everyone loses, including businesses and the public.

This book covers a broad scientific landscape, including cancer and carcinogenesis, radiation health effects, toxicology, and health risk assessment and management. The focus of the book is on general concepts and principles underlying radiological risk analysis, but the ideas developed using the radiation model are transportable to the assessment and management of chemical and other risks. The book is intended to be read by a scientifically informed audience but without specialists' knowledge. Detailed discussions of concepts and principles are presented in the early chapters in support of the major themes developed in later chapters. The book should appeal to specialists and nonspecialists interested in problems associated with assessment and management of small risks. Engineers, lawyers, policy makers, public health professionals, regulators, and scientists will be particularly interested in this book. The book contains extensive notes with references and can serve as a primary or secondary text for an advanced undergraduate or graduate course in risk analysis, occupational and environmental health, industrial hygiene and health physics.

The book is organized into ten chapters with an extensive notes and references section at the end of each chapter. Chapters 1 through 5 discuss concepts and issues in risk assessment and risk management and introduce the major themes of the book. Chapters 6 and 7 further develop the themes of a dose-based system of protection and prioritization of risk. The eighth and ninth chapters are case studies to illustrate the major themes, and the final chapter provides a summary of central ideas on dose proportion, prioritization, and precaution.

Chapter 1 introduces concepts and principles of risk assessment and risk management that are important to understanding the thematic arguments developing later in the book. Risk is a challenging concept because it has both objective, quantitative (probability) and subjective, qualitative (consequences) features. What do we mean by "acceptable risk"? How does "acceptable risk" relate to "safety"? Risk can be evaluated and managed in a number of ways. The 1983 National Academies risk-assessment model and the 1997 Presidential Commission report on risk assessment and management are discussed as straightforward approaches to risk assessment and management that have been broadly adopted by government agencies. The chapter also explores the concept of "dose" and how dose and risk are related (the dose-response function).

Chapter 2 focuses on the central element of the risk assessment process — the dose-response function. The dose-response function is used to translate measured or calculated doses into risk. Direct observations of risk (health effects) are usually obtained at doses much higher than in the occupational or environmental setting. This requires the use of theories (i.e., dose-response functions) that can be used to predict low-dose risks. The chapter explores why it is necessary to quantify risk at small doses; which predictive theories are used in risk assessment; how a particular theory is selected over biologically plausible alternatives; sources of uncertainty in dose extrapolation; and the impact of theory selection on risk management and risk communications. A key point is the need to distinguish risks based on direct observations (i.e., real risks) from risks based in theory (i.e., speculative risks).

Standards-setting organizations and various authoritative bodies have broadly adopted LNT in their cancer risk-assessment activities. Use of LNT has generated significant controversy in the scientific community. Chapter 3 explores what features of LNT make it attractive as a predictive theory; why other biologically plausible alternatives have been excluded by regulators and decision makers; and the nature of the scientific debate on application of LNT in risk assessment. The chapter also discusses appropriate and inappropriate uses of LNT particularly in radiological protection. LNT and other predictive theories are discussed in a descriptive sense with minimal mathematical treatment in keeping with the overall goal of a book written for a broad audience.

Use of predictive theories to estimate risk at low doses involves significant uncertainties that impact how risk information is used in risk-management decisions and is communicated. Chapter 4 explores key issues in risk uncertainty. At the heart of the problem are statistical and methodological limitations in measuring risk. Risks encountered in environmental and occupational settings cannot be measured directly. Use of a dose-based system of protection is introduced as a strategy to circumvent the risk-uncertainty problem.

Chapter 5 discusses risk management. The goal of risk management is risk reduction. Once a risk assessment has been completed a decision is needed concerning what to do about the risk. This chapter discusses technical and social management triggers and general management strategies. Discussions focus on ALARA (as low as reasonably achievable) and precautionary approaches. Whatever risk-management approaches are used, decision makers and risk managers must be aware of the possibility of risk–risk trade-offs and unintended consequences. Methods to reduce a target risk may result in the emergence of more serious countervailing risks.

In Chapter 6 the fascinating problem of ranking risks is discussed. The decision to manage a particular risk is complex and is anchored in how risk is prioritized by the public. Some risks are more important than others in terms of their impact on the public health, but public prioritization of risks may not reflect this. The public tends to focus on minor risks that may be uncontrollable by the public and involuntarily imposed. Major risk factors that are often voluntary and controllable have significant impact on the public health but remain essentially unregulated. For instance, cigarette smoking and certain dietary factors are responsible for about two-thirds of all cancers; other factors such as air pollutants and pesticides contribute only a few percent to the cancer burden. Yet, societal concerns (as reflected in the nature of environmental regulations) focus on the minor factors.

Chapter 7 argues for a change in the way we manage health risks. The case is made for a dose-based system of protection that is preferable to a risk-based one. The public has difficulty comprehending small risks. Risks cannot be combined because risks are frequently different and agents may interact in ways that preclude simple addition of risks. Dose is a concept readily understood by the public, and a dose-based system avoids the requirement of a predictive theory to translate dose into risk. Doses can be measured at tiny levels that are orders of magnitude below levels necessary to demonstrate health risks. A dose-based framework uses a system of dose proportions (the ratio of the measured dose to an appropriate reference dose) that can be easily calibrated without reference to health risks. A significant advantage of a dose-based system is that management effectiveness can be readily quantified by measured dose reductions.

Chapter 8 is a case study on domestic radon and lung cancer. Government management of risks of radon exposure in homes illustrates several principles developed throughout this book, including risk prioritization, cancer risk uncertainties, balancing costs and benefits, and public communication of health risks. Health risks of residential radon have been an issue at the forefront of radiological protection for more than two decades. The EPA estimates that approximately 20,000 lung cancer deaths occur annually as a result of exposure to radon gas in homes. Radon is radioactive; when inhaled, radiation from the decay of radon may damage sensitive lung cells leading to cancer. If the EPA's estimates are right, residential radon exposure is a serious public health problem (second only to cigarette smoking as a cause of lung cancer).

Chapter 9 is a case study on the health risks of cell phones. The case is a prime example of the misuse of the precautionary principle. The government of the United Kingdom (U.K.) recommends a ban on cell phone use by children based on scientific data that suggests no health risk. The U.K. debacle illustrates how risk management can run amok and how prioritization of risks can be distorted easily. Cell phone technology is available throughout the world, and it is interesting to compare cell phone risk perceptions on a national and regional basis. In spite of exhaustive study, statistically significant health risks from cell phones (particularly cancer) have yet to be identified. The cell phone case clearly illustrates the significant problem of statistical uncertainties at low doses (in this case of radiofrequency electromagnetic radiation) and the difficulties in interpreting scientific data on marginal risks.

The major themes developed in this book are summarized in the final chapter. Dose proportion is a rational solution to the practical problems of the current risk-based system of protection. Prioritization of risk is important to optimize allocation of limited public health resources. Risk-management decisions should not be made without comparison to other relevant risks. Everyone is exposed to a spectrum of risks all of the time. It is inappropriate to consider any single risk without recognizing the presence of other risks. By analyzing risks in this way, risks can be appropriately prioritized. During the past 50 years dose limits in radiological protection have continued to decrease. It is unclear how much further the drive to zero tolerance will proceed. Precautionary approaches to risk management are costly with questionable benefits. Risk-management strategies must focus on balancing costs against benefits. In radiological protection there is little justification for invoking the precautionary principle.

I have written this book because I am concerned about the extraordinary attention paid to small risks and the enormous costs to control them. We fear the wrong things and pay inadequate attention to risks that really matter. I have been thinking about this problem for many years in the context of my own work on the public health impacts of low levels of ionizing radiation. Much of my thinking has been shaped by lectures given on the topic and discussions with numerous colleagues and students. I believe that small radiation doses typical of most diagnostic x-ray procedures such as chest x-rays and mammograms are innocuous. They do not cause any measurable health effects. If we did nothing about such doses there would be no discernible impact on the public health. This view is not shared by everyone; I respect the alternative perspective that even if we cannot measure small risks we should do something about them. I would agree except that disproportionate attention is paid to managing already small risks when further reductions in large risks would have a greater impact on population health. Ionizing radiation is a weak carcinogen and a minor contributor to the U.S. cancer burden. Just because we have the technological capacity to reduce doses does not mean that we should do so. If the risk is very small to begin with, there is little benefit to be gained by reducing the risk to even smaller levels.

I am unabashedly utilitarian. This does not mean that I marginalize management of small risks. But the greatest benefit to the largest number of people will be achieved by directing resources to the control of large risks. Reduction in large risks has a greater impact on the public health than the same percentage reduction in small risks. I am concerned when huge sums of money are spent to control minor risks that are not likely to benefit public health. The characterization of transuranic waste (discussed in Chapter 6) and the Alar ban in the apple-growing industry are prime examples.

The public needs to rethink how resources are allocated. The goal of public health protection should be to protect everyone. In theory the only way to accomplish this is to take a strict precautionary approach and eliminate all technological risks by banning technologies. This is an ill-conceived approach and will do more societal harm than good because important technologies that have significant societal benefits may be entirely eliminated or significantly curtailed. In reality, we cannot protect everyone all of the time because resources are insufficient even in the wealthiest nations. Hard choices must be made, and what is needed is a balanced perspective on risk management where efforts to manage small risks are weighed against the need to control large risks. The current regulatory program in the U.S. suggests the opposite. We allocate substantial resources to manage risks that pose little threat to the public health and pay inadequate attention to larger risks (smoking, diet, lack of exercise) that contribute significantly to major disease burden.

Policy and regulatory decision making is a complex process. Social, economic, and political factors guide decision making, but the process must start with a solid scientific and technological foundation. This book argues that the quality of the science is a key to good decision making. Good science includes peer-reviewed, reproducible results and careful consideration of uncertainties. Policy decisions are indefensible when scientific knowledge is marginalized to support a particular economic or political objective. Moreover, use of faulty scientific data or overinterpreted scientific data also makes for questionable decisions.

How science is used or not used is a significant source of tension in the policy-science arena. Science is often ignored in important societal decisions even in the face of increasing public concern that decision making be based on sound science. The EPA remains under steady public pressure to incorporate defensible scientific approaches in its regulatory decision making. Part of the problem is that decision making is driven by a variety of nonscientific, adversarial, and special-interests factors. Science helps to inform choices but it is only one component in a large array of values and choices. The key to good decision making is determining what scientific information should be included and excluded, and how science should fit into the broad decision framework that also includes stakeholder concerns and economic, political, and social interests.

Prioritizing risks and allocating resources to manage risks are challenging processes at the science–policy interface. This book explores the dynamic interaction of science and policy in decision making, using assessment and control of ionizing radiation as a model system. Risks of radiation exposure are well known, and regulatory controls to limit exposures in environmental and occupational settings have been firmly established. Although the focus is on radiation, the ideas and concepts developed in the book may be broadly applied to other noxious agents including chemical carcinogens.

A daunting challenge for the U.S. public health establishment is managing major disease-causing factors such as smoking and diet because control is a question of individual behavioral modification. Small technological risks (e.g., contaminants in drinking water) require institutional controls, and management costs are governed by what can be reasonably regulated. Personal human behaviors are difficult to control by government intervention, particularly if viewed as a challenge to individual freedoms.

Etiology of major diseases including cardiovascular diseases and cancer is multifactorial. Individual disease risk profiles are combinations of exposure to disease-causing agents that can be controlled by social regulations, and individual risky behaviors that can be controlled by personal behavioral modification. Controlling known environmental risks through responsible regulation is an important strategy to protect the public health, but the major strategy should be personal risk management. Smoking cessation, proper caloric intake and dietary control, and appropriate exercise are the most effective means for reducing risks of cancer, heart disease, and diabetes for most people. A smoker living in a home with slightly elevated radon levels is much better off quitting smoking than spending money on home radon remediation to reduce lung cancer risks. We need to think globally about health risks but act individually.

Abbreviations and Acronyms

AIDS	acquired immune deficiency syndrome
ALARA	as low as reasonably achievable
AS	assigned share
AT	ataxia telangiectasia
Bq	becquerel
BAT	best available technology
BEIR	Biological Effects of Ionizing Radiation Committees of the U.S. National Academies
BRC	below regulatory concern
BSE	bovine spongiform encephalopathy
CDC	U.S. Centers for Disease Control and Prevention
CEC	Commission of the European Communities
CERCLA	Comprehensive Environmental Response, Compensation and Liability Act of 1980
CT	computerized tomography
DDREF	dose and dose rate effectiveness factor
DDT	dichloro-diphenyl-trichloroethane
DOE	U.S. Department of Energy
DNA	deoxyribonucleic acid
EC	European Commission
EDB	ethylene dibromide
EEOICPA	Energy Employees Occupational Illness Compensation Program Act of 2000
EIS	environmental impact statement
EPA	U.S. Environmental Protection Agency
EU	European Union
FACA	Federal Advisory Committee Act
FDA	U.S. Food and Drug Administration
FOIA	Freedom of Information Act
Gy	gray
HHS	U.S. Department of Health and Human Services
Hz	hertz
IAEA	International Atomic Energy Agency
IARC	International Agency for Research on Cancer
ICRP	International Commission on Radiological Protection
LLRWPA	Low Level Radioactive Waste Policy Act of 1980
LNT	linear no-threshold theory
LSS	life span study
MCL	maximum contaminant level
MCLG	maximum contaminant level goal
MHz	megahertz
MPF	modern pit facility
MQSA	Mammography Quality Standards Act of 1972

mGy	milligray
mSv	millisievert
NAAQS	National Ambient Air Quality Standards
NASA	National Aeronautics and Space Administration
NCRP	National Council on Radiation Protection and Measurements
NIH	National Institutes of Health
NNSA	National Nuclear Security Administration
NRC	National Research Council of the National Academies
NRDC	Natural Resources Defense Council
NRPB	National Radiological Protection Board (United Kingdom)
NSF	National Science Foundation
NTS	Nevada Test Site
OSCC	Oxford Survey of Childhood Cancers
OSHA	Occupational Safety and Health Administration
PAG	protective action guide
PCBs	polychlorinated biphenyls
ppb	parts per billion
ppm	parts per million
RF	radiofrequency
SEER	Surveillance, Epidemiology and End Results cancer database
SNP	single nucleotide polymorphism
Sv	sievert
TMI	Three Mile Island nuclear power plant
U.K.	United Kingdom
UNSCEAR	United Nations Scientific Committee on the Effects of Atomic Radiation
U.S.	United States
US NRC	U.S. Nuclear Regulatory Commission
WIPP	waste isolation pilot plant
XP	xeroderma pigmentosum

List of Figures

1.1 The National Research Council risk-assessment process
1.2 Correlation or causation?
1.3 Sources of radiation exposure
2.1 The dynamic relation between data and theory
2.2 Possible shapes of dose-response curves in risk assessment
2.3 Sublinear dose response
2.4 Supralinear dose response
2.5 Hormesis dose response
2.6 Threshold dose response
3.1 Linear no-threshold theory
3.2 Target model of radiation action
3.3 Cancer is a multistage process
4.1 Population size needed to detect risk
4.2 Decisions under uncertainty
5.1 Theoretical distribution of radiation response in a human population
5.2 Theoretical safety performance curve
6.1 Economic costs of risk reduction
7.1 Dose limits are unrelated to cancer risks
8.1 Distribution of radon in homes
8.2 Lung cancer risks and radon
8.3 Economic costs of risk reduction
9.1 The electromagnetic spectrum
9.2 Cell phone use and brain cancer

List of Tables

1.1 Risk: Probability and Consequence
2.1 Extrapolating Health Risks
4.1 Excess Cancer Mortality in Japanese Survivors of the Atomic Bombings at Hiroshima and Nagasaki
6.1 Ranking Cancer Risks
6.2 Influences on Prioritization
6.3 Leading Causes of Death in the U.S.
6.4 Estimated Annual Benefits and Costs of Major Federal Rules (October 1992 to September 2002)
7.1 Natural Sources of Selected Known and Suspected Human Carcinogens
7.2 Dose Proportions in Radiation Protection Decision Making
8.1 Sources of Radiation Exposure
8.2 Remediation Based on Dose Proportions
8.3 Risk Comparisons
10.1 Dose Proportions for Selected Environmental, Occupational, and Accidental Exposures
10.2 Dose Proportions for Selected Medical Exposures

1 Risky Business

To live life means to take risks. In fact, a risk-free existence is not possible. Everyone is exposed to health risks from the moment of conception until death. The greatest risk in life is being born. Being born a normal, healthy child is only a 50-50 proposition.[1] After that, each of us is exposed to risks that may lead to a spectrum of outcomes, including minor injury, serious illness requiring immediate medical attention, and death. There are inherent risks in each and every choice we make. The kinds of risks we take depend on the activities we engage in and the technologies we use. The key to living a good and healthy life is not to try to eliminate all risks entirely but to minimize risks that we can control. All automobile risks cannot be controlled by a driver, but risk-preventive practices such as wearing seat belts can be the difference between a minor injury and death.

No matter what you do and how careful you are, you cannot entirely eliminate risks of certain diseases or injuries. However, living a healthy lifestyle can reduce some risks. For example, risks for cardiovascular diseases are known to be influenced by dietary fat and high blood pressure. By enacting behavioral modifications related to diet and exercise, and taking appropriate prescription drugs, we can significantly reduce the risk of heart attack.[2] Some people live riskier lives than others because they smoke, maintain poor diet and nutrition, drink excessive amounts of alcohol, or use or abuse recreational or prescription drugs. Of course, one can never know whether a risk-prevention strategy worked for a particular individual. But we do know from population health studies that a number of risk-prevention strategies (such as wearing seat belts and reducing or eliminating cigarette smoking) have an overall public health benefit.

Actually, risk is not necessarily a bad thing. Certain technologies, such as medical x-rays, have obvious and significant individual or societal benefits that clearly outweigh any risks. To be understood, risks must be put in the proper perspective. Isolating risks without due consideration of countervailing risks or benefits leads to misperceptions and hinders decision making. For instance, the usual environmental and public opposition to radioactive waste disposal sites is based on the idea that all radiation is bad. This is a major contributor to the high disposal cost burden borne by all nuclear sectors, including the medical community.

Risks are not entirely avoidable. In spite of reasonable precautions, accidents and other risks still occur. Risks associated with automobile travel, airline travel, natural disasters such as forest fires or tornadoes, and industrial pollution are examples of consequences that are not entirely controllable by individuals. For many everyday activities, such as grocery shopping, brushing teeth, or climbing stairs, we do not think very much about possible risks.[3] For other activities such as automobile travel, smoking, or exposure to air pollution, most of us realize that some risk is involved.

This book focuses on agents that may cause cancer. Chemical carcinogens are found in air, water, and foods; physical carcinogens include ultraviolet light and medical x-rays. Why carcinogens? Because cancer is an important public health concern (it is the second leading cause of death after heart disease and stroke), and a number of environmental factors have been identified that may increase the risk of cancer. Typical carcinogen exposures are associated with small risks. Cancer-causing substances in foods, in the air, in water, and simple medical diagnostic x-rays are examples of agents that preoccupy us as health hazards even though the health risks either have not been demonstrated or are too small to be reliably measured.[4] The question of why people focus on small incidental risks and ignore or discount larger, more significant risks has been the subject of intense study. Risk perception is a complex psychosocial phenomenon. Factors such as catastrophic potential, individual control of risk, and knowledge of risk are known to be important determinants of how people view risk. Chapter 6 focuses on how individuals perceive risk and how the public prioritizes risk. This is not to say that attention is not spent on significant risks such as smoking or automobile travel. Health warnings on cigarette packs and legal requirements to use seat belt restraints in cars are examples of successful efforts to reduce risks. The problem is that disproportionate sums of money are spent managing small risks that have little, if any, impact on public health.

Although "risk" is a term that is used frequently, it is not easily definable and means different things to different people. How people feel about risks, whether or not an activity is risky, often discounts the numerical size of the risk. Risks are poorly understood by the public, and numerical expressions of risk often have little meaning. Does a risk of 1:10,000 really seem 100 times more risky than a risk of 1 in 1 million? We tend to overvalue lottery tickets and overestimate risks of airline travel but underestimate other risks such as automobile travel. We allocate limited resources to manage risks that pose little if any health concerns and ignore larger risks that have a significant health impact. The public is concerned about small risks, particularly those associated with technologies. Most technology-based risks are not well understood by the public. Risk control resides with corporate owners and government with little public input. Without any control, the public is at the mercy of business and government, and as a consequence, risk perceptions are greatly magnified. Personal behaviors such as smoking and dietary habits pose potentially significant health risks (e.g., lung cancer, cardiovascular disease, colon cancer), but they are controllable at the individual level. Yet society focuses on technologies like nuclear power operations that individuals cannot control directly but nonetheless pose an insignificant health threat under normal operating conditions.

Exposure to low levels of ionizing radiation and chemicals is associated with a special type of risk. These risks are primarily linked to cancer and are very small and extremely difficult to measure in occupational and environmental settings. Cancer is perhaps the most feared of human diseases. Familiarity with cancer and the idea that cancer is a death sentence leads to societal distortion of risk, even though the actual exposure only adds a very slight increase to the already large cancer burden.

This chapter introduces the reader to several important definitions and concepts concerning risk. Risk, risk assessment, risk management, risk communication, and safety will be introduced and briefly discussed. The discussion serves as introductory material for more detailed discussions in later chapters. The focus of the discussions will be on small risks from exposure to hazardous chemical and physical agents. Small risks and how they are assessed, managed, and communicated pose serious economic, political, scientific, and social problems.

MORE THAN A NUMBER

Risk is defined for the purposes of this book as the probability or chance of loss or adverse outcome (e.g., disease, injury, death, economic loss). Risk is the probability that an adverse effect will occur in an individual or population exposed to the hazard (i.e., an agent or activity that has the potential or capacity to cause a particular kind of adverse effect). Accordingly, risk has both quantitative and qualitative components. The magnitude of the risk is determined by its probability of occurrence and by the severity of the outcome. Risks may have different magnitudes, although their probabilities of occurrence are the same because the severities of consequences differ. For example, driving a car about 60 miles or flying in an airplane 1,000 miles have about the same probability (1 in 1 million) of causing a fatality, but airline travel is considered the greater risk.[5] Why people perceive airline travel as more dangerous is complex, but certain dimensions of risk have now been identified that explain the difference. A key factor appears to be catastrophic potential. Compared to cars, airline accidents have a very high catastrophic potential (i.e., many deaths from a single event).

Table 1.1 shows a selected list of activities illustrating different combinations of probabilities and consequences as determinants of risk.[6] In the case of alcohol consumption, just a few drinks can make some people drunk: operating an automobile under such circumstances is dangerous to the driver and others. For artificial sweeteners (e.g., saccharin), toxic levels are not reached even when used excessively.

TABLE 1.1
Risk: Probability and Consequence

Activity	Probability of Exposure to Hazardous Levels	Consequences (severity of injury to people)
Alcohol consumption	High	High
Artificial sweeteners	Very low	Very low
Caffeine consumption	Moderate	Low
Tobacco consumption	Moderate	High
Nuclear power plant accident	Very low	High
Solar radiation causing skin cancer	Very high	Moderate

Source: Entries modified from Ropeik, D. and Gray, G., *Risk*: *A Practical Guide for Deciding What's Really Safe and What's Really Dangerous in the World around You*, Houghton Mifflin, New York, 2002.

Although the consequences of nuclear power plant accidents may be severe, the probability of their occurrence is low.[7] Many people get a "buzz" from caffeinated beverages, and some may lose sleep at night. However, there are no documented serious consequences to health. Individuals smoking more than 1 to 2 packs per day have a very high risk for cancer and heart disease. Tobacco use is a major preventable cause of disease and death in the U.S. and accounts for almost half a million deaths per year.[8] The probability of sunburning can be quite high in places such as Arizona and Australia, even when one is outside for a short time. Non-melanoma skin cancers are the most common cancers, and incidence is highly correlated with sun exposure.

Risk is usually expressed as a number, frequency, or percentage of adverse outcomes (e.g., disease incidence or mortality) per number of events or dose of agent. Accordingly, two things need to be known to calculate risk — a metric quantifying outcomes (i.e., number of deaths, number of injuries, economic losses), and the amount (i.e., dose) of the agent. Acquiring this information depends on a number of factors, including the availability of direct observations or experience and specificity and temporal coupling of events and outcomes.

The risk of a fatal automobile accident is a straightforward example of how risks can be measured. Automobile accident records are readily available; traffic fatality data can also be obtained easily. The following serves as an example of a risk calculation. Let us assume the following: There are 15 million car accidents annually in the U.S. and 1 out of 300 of these accidents results in fatalities. The U.S. population is 300 million and Americans live on average 70 years.[9]

Population annual risk: 15 million car accidents × (1 fatality/300 accidents) = 50,000 traffic fatalities.

Individual annual risk: 50,000 fatalities/300 million population = 0.0002 traffic fatalities per person per year = 1 in 5,000 (probability that a person will be killed in a traffic accident per year).

Individual lifetime risk: 0.0002 traffic fatalities per person per year × 70 years = 0.014 traffic fatalities per person in a lifetime (1 in 70 persons).

Risks can be easily misinterpreted because they can be expressed in a number of ways. To be meaningful, the reference population must be clearly identified in the risk expression. Does the risk refer to the entire U.S. population or only to males over 55 years of age? Comparisons are valid only when similar risks are considered. Annual risks cannot be compared to lifetime risks; individual risks are different from population risks.

What makes this calculation straightforward is the fact that accidents and fatalities are directly observable. We can count up the number of injured or dead bodies and the number of smashed cars. Furthermore, a causal link can be readily established between the accident and the fatality. Injuries can be linked directly to the accident. Traffic fatalities usually occur immediately following the accident or soon thereafter. There is hardly an opportunity for intercurrent disease or another adverse event to occur between the time of the accident and the clinical appearance of disease or injury. Similar kinds of calculations can be made for other directly observable outcomes for which causality has been established, including alcohol-related traffic accidents and economic losses and declining stock prices.

For many kinds of risks, including the risks we are focusing on in this book, consequences of activities or agent exposure may be difficult or impossible to measure directly because the probability of occurrence is very small, there has been little if any direct experience, and a causal link has not been clearly established. When risk is based on low-frequency events, quantification is challenging, and any estimate of risk will have a high degree of uncertainty. Rare events provide little opportunity for direct observations or experience. Further, risk outcomes may be nonspecific, making it almost impossible to link outcomes to specific events or exposures. Specific outcomes may have multiple causes, making it difficult to determine the contribution of the agent in question. If the agent is neither necessary nor sufficient to cause the outcome, it may be difficult or impossible to link the risk to a given agent. If there is a significant temporal uncoupling between the exposure and effect (e.g., cancer may occur years after exposure), establishing causality may be extremely difficult. Furthermore, if the disease has a large background incidence, it is almost impossible to reliably measure outcomes caused specifically by the subject agent.

These limitations usually necessitate dose extrapolation to predict risks at small doses based on direct observations made at very high doses. The high doses are necessary in order to increase the probability of seeing the effect. Depending on the agent (radiation, industrial chemicals, pesticides, etc.), human populations exposed to high doses may be available from occupational, medical, or military exposure situations. For instance, much of what is known about the cancer-causing effects of ionizing radiation is derived from the long-term study of the Japanese atomic bomb survivors.[10] Additional human data are also available from medical exposures where patients were given high doses of radiation for treatment of cancer and noncancer diseases. But human populations exposed to high doses of many chemical agents (e.g., most chemicals regulated by the Environmental Protection Agency [EPA]) are not available. In these situations laboratory animal studies are conducted to determine the magnitude of risk at high doses. High dose experimental studies have the advantage of maintaining lower experimental costs since fewer animals are needed to demonstrate a risk if it exists. Risks derived from animal (or human) studies conducted at high doses are extrapolated downward to calculate the risks to be expected at doses typically encountered in the environmental or occupational setting. Dose extrapolation, as discussed in Chapter 2, is associated with substantial scientific uncertainty, such that inferred risks must be interpreted with great caution. The use of animal studies introduces another source of uncertainty referred to as cross-species extrapolation. In conducting animal studies, it is assumed that the effect seen in a particular animal model mirrors the effect in humans. For chemicals, this means that the selected animal model handles and metabolizes the agent in the same way a human would and produces the effect in the same way that a human would. Accordingly, there may be substantial variations in risk estimates depending on the animal model selected because of differences in handling and metabolism of the agent.

For the purposes of this book, low risk is defined as a lifetime risk of less than 1:1000.[11] Examples of low-risk events include the probability of cancer from exposure to air pollutants, and the probability of dying in an airplane crash. It should not be inferred that a risk greater than 1:1000 is a high risk. This is a working definition that provides a perspective on the magnitude of the probability of small risks. Obviously, there is some personal judgment in selecting a particular value as a defining boundary. What one person may consider as a low-risk event, another person may consider significant.[12] However, activities such as cigarette consumption, excessive alcohol consumption, or travel in an automobile at high speed would be considered high-risk events by most people. In this book we focus on risks as a result of exposure to low levels of cancer-causing agents.

SAFETY WITHOUT RISK?

How do we know that an over-the-counter medication is "safe"? Does it mean that the probability of complications or side effects from taking the drug is zero? If not, what level of risk is acceptable to establish safety? According to Webster's dictionary "safety" is defined as: *1. free from harm or risk. 2. secure from threat of danger, harm or loss. 3. affording safety from danger.* The first definition implies that safety is achieved only when risk is zero, suggesting that the only way that safety can be achieved is by eliminating the agent entirely.

Perhaps the best-known example of zero tolerance is the Delaney Clause, which was added as part of the Food, Drug, and Cosmetic Act of 1958. The Delaney Clause requires that no carcinogen (as demonstrated in either humans or animals) will be deliberately added to or found as a contaminant in food. In practice the clause applied to processed foods but not to fresh fruits or vegetables. Strict adherence to the Delaney Clause required that a single carcinogen molecule be unallowable. The Delaney Clause came into existence in part because of scientific and technological advances in measuring very small amounts of potentially harmful substances. In some ways we are victims of our own success, and the Delaney Clause created a crisis situation. As we acquired the technology to measure miniscule amounts of substances, most foods we eat would have been affected under Delaney. In August 1996, Congress passed the Food Quality Protection Act of 1996 repealing the Delaney Clause and, instead, established a less restrictive but more scientifically defensible standard of "reasonable certainty of no harm" to public health.[13]

A zero-tolerance philosophy is unrealistic and counterintuitive. Eliminating the target risk entirely may have the unintended consequence of increasing risks from activities unrelated to the one being managed. Risk-risk trade-off is a zero sum game! If any risk is unacceptable, then nothing is safe. Under Delaney, eventually every food would be banned from the dinner table because of trace quantities of pesticides and even naturally occurring substances that might cause cancer. Every activity we engage in or every agent we are exposed to has some risk associated with it. The risk may be imperceptibly small but it is not zero. The fact that we can measure small amounts of substances in food, air, and water does not mean that there is an associated health risk. If a minute amount of a substance can be detected, does it mean that it is not safe? Radiation detection has now become so sensitive that we can measure single radioactive atoms; if we can measure radiation down to single radioactive atoms, does that mean that a few atoms pose a health risk and are not safe?

When irresponsible actions are taken or public confidence in products or services is eroded, safety may be seriously questioned even if risks are small. An automobile is no longer considered safe when a drunk gets behind the wheel. The Tylenol® scare of 1982 illustrates how product tampering can quickly change public perceptions of drug safety.[14] In fall 1982, Johnson & Johnson was confronted with a crisis when seven people on Chicago's West Side died mysteriously. Authorities determined that each of the people that died had ingested an Extra-Strength Tylenol capsule laced with cyanide. The news of this incident traveled quickly and was the cause of a massive, nationwide panic. These poisonings made it necessary for Johnson & Johnson to launch a public relations program immediately, in order to save the integrity of both its product and its company. After a massive national recall program, Johnson & Johnson reintroduced Tylenol with new triple-seal packaging that made the product essentially tamper-resistant. The new product safety design and an effective public relations and mass marketing campaign reestablished public confidence in Tylenol.

These examples suggest that safety (and risk) is really a moving target. Determining safety is a subjective enterprise involving the acceptability of a risk. In some instances a product or activity may be safe and in other situations it may not be. A more practical definition of safe is the level of activity or agent exposure that is deemed to be an acceptable risk. A product or activity may be considered safe if its risks are judged to be acceptable.[15] The obvious question is: acceptable to whom? A risk may be deemed acceptable (i.e., safe) to one person but unacceptable to another. A butcher's knife may be deemed safe in the hands of an adult but unsafe in the hands of a child.

Safety is a highly relative attribute that can change from time to time and be judged differently in different contexts. An agent may be considered safe today but not safe tomorrow when we will know more about it. Knowledge of risks evolves, and so do the personal and social standards of acceptability. Some nations have different standards of safety (and levels of risk acceptability) because they can afford to be more restrictive. For example, DDT (dichloro-diphenyl-trichloroethane) is essentially banned in the U.S. because we have access to more costly pesticide alternatives. Third-world countries continue to use DDT because it is inexpensive and they cannot afford more costly options. Nuclear radiation is considered more risky today than 50 years ago because we know more about the health effects and can detect radiation with increasingly more sophisticated equipment that essentially allows us to find single radioactive atoms. If we can detect an agent, does that mean it is dangerous?

WHAT'S RISKY?

How do we determine how hazardous or risky a particular agent might be? What information do decision makers need to know to decide to pull a product off the market because it contains a substance that has a risk of causing cancer? Why did the U.S government decide to list saccharin as cancer-causing in 1977 and then 23 years later removed the carcinogen label? The process or procedure used to estimate the chance that humans will be adversely affected by a chemical or physical agent is called risk assessment. Risk assessment essentially looks at three questions: What can go wrong? How likely is it to happen? What are the consequences? The purpose of risk assessment is to provide pertinent information to risk managers, policy makers, and regulators so that the best possible decisions can be made regarding management of the risk. Risk assessments may be conducted without considering what the actual exposures may be to a population at risk. Since about 1970 the field of risk assessment has received widespread attention in both the scientific and regulatory communities.[16] Risk assessment is an important tool in remediating Superfund sites, improving safety at chemical processing plants, and improving safety at nuclear and fossil-fueled electric power-generating stations. Risk assessment, particularly involving very small exposures to hazardous agents, has a high degree of uncertainty, but conservative safety margins are built into an assessment analysis to ensure protection of the public. To make sure that all members of the population at risk are protected, risk analysis may focus on sensitive population subgroups (e.g., children and pregnant women). Protecting the vulnerable protects everyone.

In 1983 Congress funded a study conducted by the National Research Council (NRC) to strengthen the reliability and objectivity of scientific assessment that forms the basis for federal regulatory policies particularly for carcinogenic agents. The result was the NRC "Red Book," which introduced a framework for risk assessment applicable to a broad range of possible carcinogenic agents. Federal agencies that perform risk assessments are often hard-pressed to clearly and convincingly present the scientific basis for their regulatory decisions. In the recent past, decisions on saccharin, nitrates in foods, formaldehyde use in home insulations, asbestos, air pollutants, and a host of other substances have been called into question.

The NRC defines risk assessment[17] as follows:

> The characterization of the potential adverse health effects of human exposures to environmental hazards. Risk assessments include several elements: description of the potential adverse health effects based on an evaluation of results of epidemiologic, clinical, toxicologic, and environmental research; extrapolation from these results to predict the type and estimate the extent of health effects in humans under given conditions of exposure; judgments as to the number and characteristics of persons exposed at various intensities and durations; and summary judgments on the existence and overall magnitude of the public-health problem. Risk assessment also includes characterization of the uncertainties inherent in the process of inferring risk.

Risk assessment can be divided into four major elements, as illustrated in Figure 1.1. This framework provides for a systematic approach to the science of risk assessment that all government agencies can utilize. The NRC report focused on chemical carcinogens, but the framework is also applicable to ionizing radiation and other physical and biological carcinogens. In practical risk assessment, elements may be considered in parallel, and information about one element may be important in the evaluation of other elements. In many practical instances, exposure assessment reveals that the target group or population may be exposed to doses that are well below the range of data available in the dose-response assessment. Under such

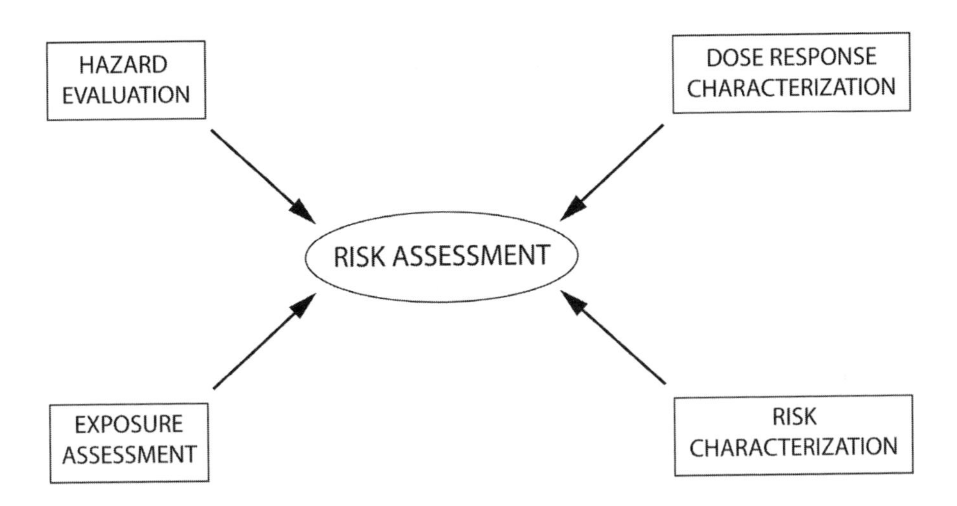

FIGURE 1.1 The National Research Council risk-assessment process. The four elements of risk assessment are not independent of one another. (From National Research Council, *Risk Assessment in the Federal Government: Managing the Process,* National Academy Press, Washington, DC, 1983.)

circumstances, dose-response assessment may have little value in the risk-assessment process because of large uncertainties associated with dose extrapolations. A good example of this involves assessing health risks from radionuclide emissions from nuclear power plants. The estimated population doses are orders of magnitude below doses that are known to increase health effects in exposed populations. The magnitude of the required dose extrapolation limits the utility of dose-response assessment as a determinant of risk. Risk uncertainties derived from plausible dose-response alternatives likely overlap. Uncertainties in risk are so large that the shape of the dose-response curve at very small doses is irrelevant. The limitations of dose-response assessment are discussed more fully in Chapter 2.

IS IT DANGEROUS?

The first step in the risk-assessment process is hazard identification, defined as the process of determining whether human exposure to an agent has the potential of causing an increase in the incidence of health effects such as cancer. Is the agent dangerous in any situation, or is it dangerous only when used under specific conditions? For example, radioactive material in powder form is likely to be less hazardous if used as a sealed source rather than as uncontained powder that is free to disperse. Hazard identification involves examining the nature and strength of the evidence of causation. Causation should be clearly distinguished from statistical association. A statistical association simply describes the behavior of one variable with respect to another. If the variables are independent of one another, then no causal relation exists.

Figure 1.2 illustrates the distinction between statistical association and causation. Lung cancer deaths can be shown to increase with increasing cigarette consumption

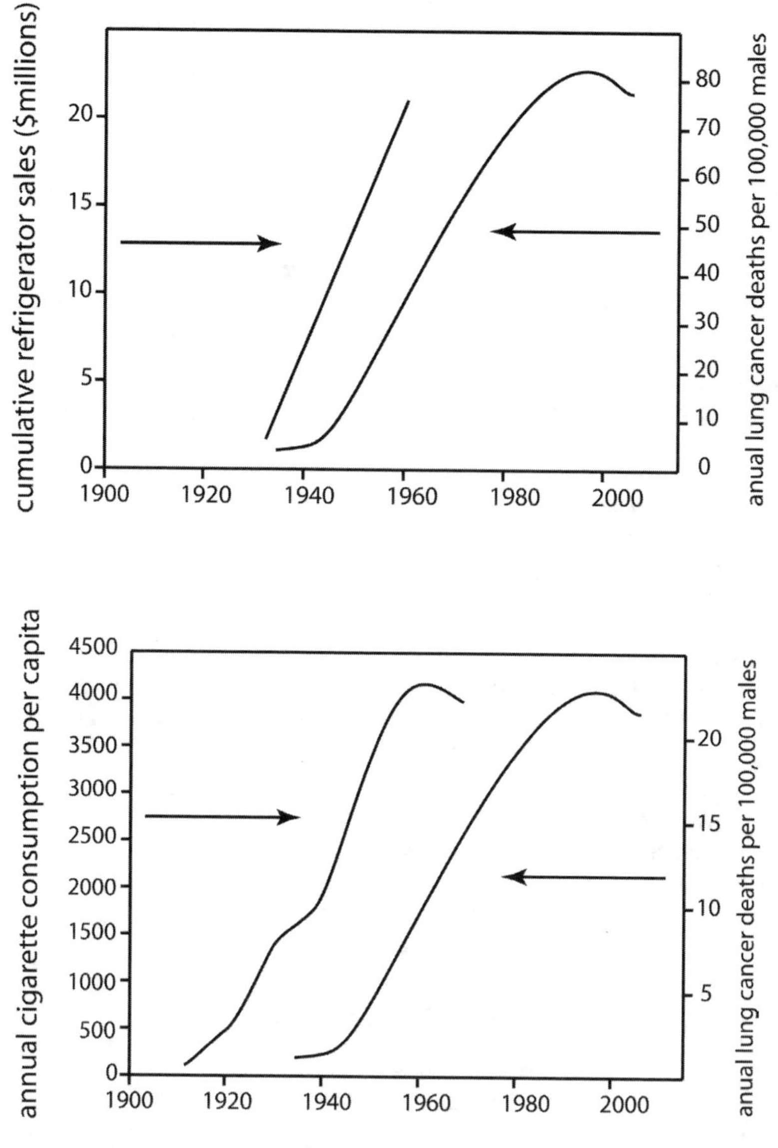

FIGURE 1.2 Correlation or causation? Refrigerator sales and cigarette smoking both show high correlation with lung cancer mortality but only cigarette smoking is a causal factor for lung cancer. Statistical correlation is only one factor in establishing causality. (From American Cancer Society, *Cancer Facts and Figures*, American Cancer Society, Inc., Atlanta, published annually; World Health Organization, *The Tobacco Atlas*, World Health Organization, Geneva, 2002.)

and with increasing sales of refrigerators. From these graphs alone, it is not possible to determine that cigarettes are a direct cause of lung cancer deaths and that refrigerator sales have nothing to do with lung cancer deaths. Both show a strong statistical correlation with lung cancer deaths. However, cigarette smoking is causal because a biologically plausible explanation can be offered. The key piece of causal evidence for cigarettes is that reduction in cigarette consumption leads to a reduction in lung cancer mortality. Thus, simply establishing a statistical association between an agent and particular disease is insufficient for a finding of causality. A.B. Hill identified a number of characteristics for the establishment of causality.[18] If causality can be reliably established, then risk of exposure to the agent can be derived from the statistical association.

However, establishing causality may be quite difficult. In reality not all of Hill's postulates can be clearly demonstrated for most causal agents. Support for causality will depend on the strength of the collective evidence. The most clear-cut case for causality is derived from situations where the subject agent is both "sufficient" and "necessary" to cause the disease. An agent is deemed "sufficient" if, in the absence of other causal factors, it is capable of producing the disease. However, this condition does not preclude the possibility of multiple "sufficient" agents. On the other hand, an agent is "necessary" if it is required to be present for the disease to occur but other factors may also be "necessary." Causality is much more difficult to establish when agents are characterized as either "necessary" or "sufficient," or neither "necessary" nor "sufficient."

What kind of evidence is used to establish whether a particular agent is a hazard? Ideally, one would like to have evidence directly available from humans since it is risk assessment in humans that we are trying to establish. However, there are very few agents for which human data are available. For human data to be of value in hazard identification and risk assessment, exposure conditions need to be well defined, the characteristics of the exposed population should be well known, and a broad range of doses to individuals should be available. Complete and accurate medical records documenting the disease of interest should be maintained. Ionizing radiation is one of the few agents for which an extensive human experience is available. Human exposure data and epidemiological studies are available from a variety of sources, including military uses of nuclear weapons, and medical uses of radiation in the diagnosis and treatment of various benign and neoplastic diseases (i.e., cancer). Unfortunately, very little human experience is available for most chemicals found in the environment or workplace.

Human evidence (if available) of a hazard is supplemented by studies in laboratory animals or other kinds of laboratory test systems. The key assumption in conducting animal studies is that disease mechanisms in laboratory animals and humans are essentially the same. Species may have different sensitivities to the particular agent, but if the agent causes cancer or some other disease in one species, it is assumed that it will also produce the same disease in humans. Predicting the behavior and health effects of a particular agent in humans from studies in laboratory animals may be highly uncertain. Cross-species extrapolation introduces an additional layer of uncertainty in the risk-assessment process. If an agent is a hazard in a laboratory model system, it may or may not be hazardous for humans. The reverse may also be true, as illustrated by the thalidomide tragedy.[19] Even if an agent is shown to be a hazard both in humans and test animals, species sensitivities may vary considerably. One species may show a response at a low dose for the agent but much higher doses might be necessary to show the same response in another species. A given laboratory animal model may be a poor predictor of the human response because the animal may respond to doses that are substantially higher than what is typically encountered in the human situation. Tests to establish saccharin (an artificial sweetener used in soft drinks) as a carcinogen provide a good example. Saccharin-induced cancer has never been demonstrated in humans. However, in laboratory animals it has been observed when huge doses of the chemical were administered. The dose is so high that an individual would have to drink over 1,000 cans of soft drink a day to ingest the human equivalent of the saccharin needed to cause cancer in laboratory animals. Is saccharin really a human carcinogen if laboratory animals need unrealistically high doses to demonstrate a health effect?

At occupational or environmental levels of a carcinogen, the probability of induced cancer may be so small that it cannot be measured. Although hazardous in a theoretical sense (because cancers were detected at very high doses), the agent may be considered for all practical purposes nonhazardous during routine use. Regulatory agencies like the EPA have taken the position that if an agent is considered hazardous at high dose, it is also hazardous at low dose.

In addition to epidemiological studies in humans and long-term animal studies, hazard characterization also involves extensive analyses of physical-chemical properties and routes and patterns of exposure to the subject agent. Metabolic and pharmacokinetic properties and toxicological studies are also conducted in test animals.

The EPA and the World Health Organization's International Agency for Research on Cancer (IARC) use similar hazard identification schemes based on the quantity and quality of human and laboratory animal evidence available in support of carcinogenicity.[20] The classification schemes are applicable to carcinogens, but with appropriate modifications might also be used for agents causing noncancer health effects. In the EPA scheme, an agent can be placed into one of five categories depending on the type of available evidence (human and/or animal data): (1) carcinogenic in humans, (2) likely to be carcinogenic in humans, (3) suggestive evidence of carcinogenicity but not sufficient to assess human carcinogenic potential, (4) data are inadequate for an assessment of human carcinogenic potential, and (5) not likely to be carcinogenic in humans. The IARC system groups agents into four categories. In Group 1, there is sufficient epidemiological evidence that human exposure is causally linked to cancer. In Group 2A, the agent is *probably* carcinogenic if there is limited human evidence but sufficient animal evidence. An agent is considered *possibly* carcinogenic (Group 2B) if there is little human evidence and insufficient animal evidence. An agent is considered nonclassifiable (Group 3) if there is inadequate human or animal evidence to make a carcinogenicity determination. Agents placed in Group 4 are considered noncarcinogenic if there is evidence suggesting lack of carcinogenicity in humans or in animals. To date a relatively small number of physical agents, chemical agents, and various mixtures have been classified as human carcinogens, including arsenic, asbestos, benzene, radon gas (a naturally occurring radioactive gas), and vinyl chloride.

The National Toxicology Program of the U.S. Department of Health and Human Services (HHS) also rates human carcinogens using a simpler system. Agents are either *known to be human carcinogens* or are *reasonably anticipated to be human carcinogens*. The first category is equivalent to IARC Group 1 that requires sufficient evidence from human studies. The second category is a hybrid of IARC Groups 2A and 2B that require at least some animal data in support of carcinogenicity.

Clearly, there is substantial personal judgment in the hazard classification of carcinogens. Unless the evidence is substantial (i.e., solid human evidence), subjective assessment plays a major role in evaluating what human and animal data say for the purposes of classification. Mistakes have been made. Some agents have been listed as possible carcinogens and removed from the marketplace on that basis. The classification of saccharin is a good example. Saccharin was first listed as a possible human carcinogen in 1977 based on no human data and very limited animal data. A classification review by HHS resulted in delisting of saccharin as a carcinogen in 2000.[21] IARC also downgraded saccharin from possibly carcinogenic to unclassifiable.

X- and gamma rays are listed as known human carcinogens in the 11th edition of HHS's *Report on Carcinogens* because human studies show that exposure to these types of radiation causes many types of cancer, including leukemia and cancers of the thyroid, breast, and lung.[22] IARC also labeled x-rays as a human carcinogen. This is an interesting situation. There is abundant evidence that high-dose medical radiation for the treatment of human diseases (particularly cancer) is carcinogenic. However, radiation risks are much less certain in the low-dose range that includes diagnostic procedures. Should all medical x-ray procedures be labeled as cancer causing or should dose matter when labeling an agent a human carcinogen? It makes sense to label high-dose x-rays as a human carcinogen because there is epidemiological evidence to show that treatment of benign disease with large doses of x-rays increases the risk of cancer. However, there is very little evidence to show that low-dose diagnostic procedures are carcinogenic. Some procedures such as cardiac catheterization are diagnostic but involve relatively high doses. The cancer risks of radiation exposure are well documented in studies of atomic bomb survivors and for high-dose therapeutic procedures. The risks from small doses of radiation as received in a chest x-ray or dental x-ray are controversial. The risks are so small that if they exist, they are too small to measure reliably.

CAN I GET EXPOSED?

Risk assessment also includes evaluation of agent exposure. Who can get exposed? Under what circumstances can exposure occur? Are there particular occupations or workplace conditions where exposures may be problematic? How many people in a particular occupational group or in the general population will be exposed to different doses of the agent? What is the duration of exposure? Exposure assessment is the process of measuring or estimating the intensity, frequency, and duration of exposure to agents in the workplace or environment. This includes characterizing the source term. For instance, does exposure occur in particular industrial processes but not in others, or is the source naturally occurring resulting in exposure to larger segments of the population? What are the major pathways of exposure to humans?

Is the agent found in water, air, or other media? What are the concentration levels of the agent in a particular medium? What are the sources and magnitudes of uncertainties associated with exposure assessment?

Exposure to a hazardous chemical agent occurs when it comes into physical contact with the individual. This can occur by inhalation (contact with the lungs), ingestion (contact with the gastrointestinal tract), or body surface contact (contamination of the skin). The route of exposure is usually dictated by the medium in which the agent is carried. Exposure to airborne agents can occur by inhalation or surface deposition. But inhalation is not likely to result in exposure to waterborne agents.

Some nonchemical agents do not require direct contact for exposure. Ionizing radiation and other external radiation sources (e.g., cell phone towers emitting radiofrequency electromagnetic radiation) do not have to be in direct contact with the individual to affect an exposure because the radiation emanating from these sources can travel through various media (air, water) and interact with an individual.[23]

Exposure may not result in harm. If a particular health effect requires that the agent interact with a target tissue or organ, no effect would be expected unless such interaction occurred. Exposure simply means that the individual came into contact with the chemical agent or radiation. What is important for risk assessment is the dose of the agent in the target tissue or organ. The dose to a particular tissue determines the probability of adverse health effects in that tissue. If a particular agent is nephrotoxic, the agent will not cause damage provided there is no dose to the kidneys. For a dose to occur, the agent must penetrate the skin, cross the lung or gastrointestinal barrier, and accumulate in the target tissue (usually via the general circulation). The dose usually represents some fraction of the exposure since not all of the agent ultimately finds its way to the target tissue.

Exposure assessment can be direct (e.g., personal exposure monitoring), or can be estimated indirectly using measurement methods in various environmental media and patterns of human activity. Computer models are used to estimate doses to specific tissues and organs based on input exposure data.[24] Assessing the magnitude of exposure in a population may be difficult. Often surrogate measures of exposure are used in the absence of reliable direct measurements. Whether exposure and doses are assessed directly or indirectly, there can be substantial uncertainty in the estimates. These uncertainties must be carefully considered in the risk-assessment process. Overly conservative assumptions in computerized models can lead to unrealistic estimates of exposure and dose.

Exposure to ionizing radiation has been well characterized. Ionizing radiation can be easily measured, and sources of natural background radiation are well known. The major source of ionizing radiation to human populations is inhalation of radon gas, accounting for about half of the total natural background exposure (Figure 1.3). Radon gas and its health effects are discussed in detail as a case study in Chapter 8. Natural background radiation (Figure. 1.3, left pie chart) provides about 82% of the total annual exposure to humans from all sources. Cosmic radiation comes from deep outer space; external terrestrial radiation comes from radionuclides in the environment; radon gas is the principal source of inhaled radionuclides; radioactive potassium in foods (e.g., bananas) is the principal source of exposure via ingestion. Medical radiation, including diagnostic x-rays, is the major anthropogenic source

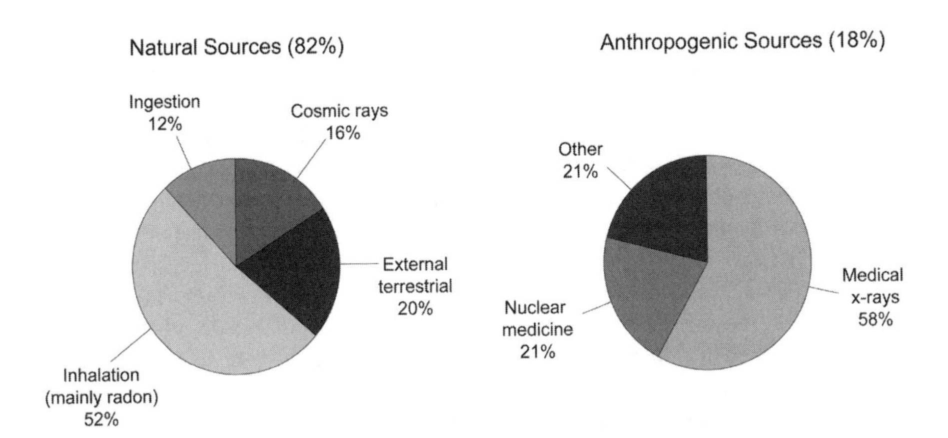

FIGURE 1.3 Sources of radiation exposure. Humans are exposed to a variety of natural and anthropogenic sources. (From National Council on Radiation Protection and Measurements, *Ionizing Radiation Exposure of the Population in the United States,* NCRP Report 93, National Council on Radiation Protection and Measurements, Bethesda, MD, 1987.)

of exposure. Anthropogenic sources (Figure 1.3, right pie chart) account for about 18% of the total annual exposure. More than three-fourths of the U.S. population receives medical or dental x-rays annually. Radiation exposure from nuclear power plant operations and the nuclear fuel cycle, including radioactive waste disposal, account for less than 1% of the total annual exposure.

The left pie chart should be interpreted with caution because radon cannot be easily compared with the other sources of natural background radiation. Radon gas exposes the lung and is linked only to lung cancer. The other sources of natural background radiation expose the whole body, and risks include lung and other cancers. To calculate the contribution of radon and other sources to the total natural background, a whole-body equivalent dose (called the effective dose) is determined for each source. Calculation of the effective dose for radon is problematic because the lung is the only important target organ and lung cancer is the only important health effect. Accordingly, radon should be considered as a separate source and not combined with other natural background sources.

CAN IT HURT ME?

Perhaps the most important step in the risk-assessment process is the determination of the shape of the dose-response curve. How does the health risk vary with exposure to the agent? Over 400 years ago, Paracelsus recognized the importance of this question when he said, "The dose makes the poison." The dose response characterizes the relationship between the dose of an agent and the incidence of an adverse health effect in a population. The dose response is important for several reasons. First, it is a key component in setting regulatory standards for control of exposure in the workplace and the environment. For carcinogens, the dose response is assumed to be linear without a threshold (i.e., there is no dose below which an induced effect or response is possible). The slope of the linear dose response is a measure of the risk per unit dose. Dose response is used to extrapolate from high doses where health effects may be readily observable down to lower doses that may be relevant in the occupational and environmental setting. Second, health effects can be predicted from the dose response if population doses are known. This is valuable information for risk managers who must decide what strategies and what resources should be allocated to manage a particular risk. Third, the shape of the dose response may provide some insight into underlying mechanisms of health effects. The linear, no-threshold dose response (LNT) used for carcinogens suggests that any dose of the agent may be harmful. If the dose response is characterized by a threshold (a dose below which no agent-induced health effects are observed), then a certain dose level must be exceeded before health effects are observed.

Ideally dose response should be based on human data, but as discussed earlier, such data are often either nonexistent or are very limited. Animal studies are conducted to predict what the dose response might look like in humans. Selection of the appropriate animal model must be carefully considered since in extrapolating the data to humans there should be a reasonable assumption that the health effects observed in the animal model are likely to mimic the health effects in humans. To be of value, animal studies must be carefully controlled, including the doses evaluated, route of administration, rate of dose delivery, and the duration of the experiment to measure health effects. A serious limitation of animal studies is that high doses are often needed to demonstrate effects. These doses may be unrealistic when compared to doses typically encountered by humans in environmental or occupational exposure settings.

Dose-response studies conducted in animals can provide important information in evaluating the dose response in humans. Animal results may elucidate mechanisms of action of the agent. Data may also be valuable in determining whether a dose threshold is likely to exist or whether there are any effects over the dose range of interest. Animal studies may also provide evidence for exclusion of certain biologically plausible dose-response alternatives. For instance, if a threshold dose was clearly present, this would be *prima facie* evidence that a LNT dose response is probably incorrect.

An important theme developed in this book is the limitation of dose response to predict risks at very small doses, particularly when very large dose extrapolations are involved. The saccharin case discussed earlier is a good example. At low doses, estimates of risk can be very uncertain and their value in decision making and risk management is questionable. The dose-response question is discussed more completely in Chapters 2 and 3. In Chapter 2 several dose-response alternatives are discussed. Federal agencies in the U.S. rely on LNT theory in regulating physical and chemical carcinogens. LNT is discussed in detail in Chapter 3.

WHAT ARE THE RISKS?

The final stage of the risk assessment is risk characterization. This is the process of estimating the probability of health effects under various conditions of human exposure. Risks are generally calculated by multiplying the measured or calculated dose to the individual or population by the risk per unit dose as determined from the dose-response function. Clearly, the magnitude of the risk estimate will depend on the underlying dose-response function used to determine the risk per unit dose. Federal regulatory agencies use LNT in part because it gives conservative estimates of the risk per unit dose. As the final step in the risk-assessment process, risk characterization must include careful consideration of uncertainties in risk and identification of key sources of uncertainty. To do otherwise would imply that risks are known with a degree of certainty that is not borne out by the quality of the data. For the purposes of this discussion, uncertainty refers to a quantity that has a true but unknown value. An example would be the effectiveness of a particular agent to cause cancer. Variability, on the other hand, refers to the different values that a particular quantity might take on because of differences among individuals, locations, time, and so forth. An example might be height or body weight. If dose response is based on animal data, cross-species extrapolation may be a key source of uncertainty. If large doses are needed to demonstrate health effects in animals, substantial uncertainties may derive from dose extrapolation to the lower doses typically encountered in human exposure scenarios. Uncertainties in the exposure assessment and in the dose-response assessment are propagated in the final determination of risk.

Risk managers, policy makers, regulators, and others interested in the risk-assessment process characterize risk to determine the health detriment (including number of disease cases and number of deaths) in a population of interest. Risks may also be estimated in vulnerable populations and in highly exposed groups to characterize the maximally exposed individual. Risk characterization is also important in identifying dominant or relatively important exposure pathways. Knowledge about these pathways is important in developing strategies to reduce exposure.

DAMAGE CONTROL

The purpose of risk assessment is to provide pertinent information to risk managers so that decisions can be made to reduce risk where possible. Risk management lays out alternative strategies to reduce the risks. Risk management is separate from the risk-assessment process and is conducted by individuals who are normally not part of the risk-assessment process. The National Research Council defines risk management[25] as follows:

> The process of evaluating alternative regulatory actions and selecting among them. Risk management, which is carried out by regulatory agencies under various legislative mandates, is an agency decision-making process that entails consideration of political, social, economic, and engineering information with risk-related information to develop, analyze, and compare regulatory options and to select the appropriate regulatory response to a potential chronic health hazard. The selection process necessarily requires the use of value judgments on such issues as the acceptability of risk and the reasonableness of costs of controls.

Following assessment of risk, risk management explores risk valuation, optional management strategies, and socioeconomic implications. Managers are concerned about the acceptability of risk and what needs to be done to reduce the risk: At what point in the risk-management process is the residual risk considered acceptable? To whom is it acceptable? What strategies are available to reduce the risks? What are the economic and social costs of risk reduction?

Let's consider a food additive that is a known or is a probable human carcinogen as an example of what a risk manager must consider in the process of controlling the risk. What are the economic and health costs of banning the additive entirely? What social benefits might be lost? Suppose the food additive is an effective artificial sweetener that diabetics (who need to control intake of sugar) rely on to sweeten sodas and coffee? If the sweetener is banned, what options are available to diabetics? Does the ban result in the emergence of other risks (e.g., other artificial sweeteners are not as effective and diabetics resort to the use of sugar that could have long-term health risks)? If the decision is to ban the food additive, what is the cost of regulatory compliance? What is the cost of developing appropriate tests and implementing such tests to ensure that the additive is no longer present in food products? Answers to these questions cannot always be put into quantitative terms and cannot be reduced to binary alternatives. A lot of assumptions, guesswork, and subjective assessment are involved.

A 1997 report from the Presidential/Congressional Commission on Risk Assessment and Risk Management extends the ideas developed in the NRC 1983 Red Book to include a comprehensive framework for risk management.[26] Although the Red Book has proven to be a useful guide for risk-assessment practices, the report does not fully explore how risks can be managed effectively. The Red Book also fails to consider risks in a comprehensive framework; it views target risks independently of other risks, and risks are evaluated on an agent-by-agent basis. In reality people are exposed to multiple agents simultaneously in both occupational and environmental settings.

The Commission proposed a distinctive risk-management approach to guide risk assessment, risk communication, and risk reduction. Historically there has been little progress in developing a generally acceptable framework for making risk-management decisions. In the Presidential/Congressional Commission report, a comprehensive and systematic framework has been developed that focuses on various contaminants, media, and sources of exposure, as well as public values, perceptions, and ethics. The risk-management framework is characterized by six stages: (1) formulating the risk problem, (2) analyzing the risk, (3) defining risk-management options, (4) making sound risk-management decisions, (5) implementing risk-management decisions, and (6) evaluating the effectiveness of actions taken. A key feature of this framework is stakeholder involvement. To be effective, risk management must be contextualized. Unless interested and impacted individuals and groups have a role in decision making, subsequent buy-in of a specific risk-management program may be difficult.[27]

The framework requires that specific problems be placed in a broader context of public health or environmental health and that the interdependence of related risks be considered in risk analysis and risk-management options. Risks cannot be evaluated in isolation but must be considered in the context of other competing risks. Risk-reduction options are evaluated in terms of their benefits, costs, and social, cultural, ethical, political, and legal dimensions.

The framework developed in the Presidential/Congressional Commission report has been used by the U.S. Nuclear Regulatory Commission (U.S. NRC) to develop in part its risk-informed approach to regulatory decision making. Risk-informed decision making is a philosophy whereby risk insights are considered with other factors to address design and operational issues commensurate with their importance to health and safety. A "risk-informed" approach is characterized by (1) allowing consideration of a spectrum of potential challenges to safety, (2) providing a means for prioritizing these challenges based on risk significance, operating experience, or engineering judgment, (3) facilitating consideration of resources to defend against these challenges, (4) explicitly identifying and quantifying sources of uncertainty, and (5) providing a means to test the sensitivity of the results to key assumptions thus leading to better decision making. A risk-informed regulatory approach can be used to reduce unnecessary conservatism in deterministic approaches, or can be used to identify areas with insufficient conservatism and provide the bases for additional requirements or regulatory actions. The risk-informed approach has been incorporated by the U.S. NRC in its reactor safety program and in its regulatory framework for uses of radioactive material in medicine.[28]

The U.S. NRC's risk-informed performance-based regulatory approach recognizes that activities do not pose the same risks and should not be regulated or controlled to the same level of detail. Prioritization identifies activities that pose the greatest safety hazards. A risk-informed philosophy results in less prescriptive regulation for the control of minor risks. As an example, NRC has shifted its focus away from detailed reviews of compliance records during license inspections toward performance issues such as prevention of errors that might result in incorrect radiation doses to patients.[29]

In Chapter 5 the process of risk management as a health, safety, and socioeconomic enterprise is discussed in detail. The focus is on ALARA (as low as reasonably achievable) and the precautionary principle as risk-management strategies. Historically, ALARA has been the driving philosophy in radiation protection practice. But precautionary approaches whereby the goal is to drive risks to zero have become increasingly more visible.

PERCEPTION IS REALITY

An integral part of the risk-assessment and risk-management exercise is framing and communicating risks. In some ways this represents the most challenging part of the risk problem. If expressed improperly, risk information can result in misunderstandings and incorrect messages. For instance, in discussing side effects of medications, very different meanings may be extracted depending on whether the risks are expressed as a percentage or as a frequency. Some antidepressant drugs are associated with a 30% to 50% chance of sexual dysfunction, and unless the patient has an understanding of the risks, he or she may be reluctant to take the medication. When expressed as a percentage some patients may interpret risk to mean that in 30% to 50% of their sexual encounters something would go wrong. When the risk is expressed as a frequency, 3 to 5 people out of 10 will experience side effects, the meaning is clearer, and patients may be less anxious about taking the drug. The risks are equivalent whether expressed as a percentage or as a frequency but framing the risk in terms of frequencies may be more understandable.[30]

Risk assessment is primarily carried out by scientists who may be quite detached from the real-world activities that involve the risks they are studying. They often express risks in ways that are not understandable by the public. In addition to assessing risk, scientists have a responsibility to distill scientific and technical information into a package that can be readily comprehended by risk managers and the public. Risk managers similarly must be able to effectively communicate highly technical information in easily understandable terms for policy makers and the public. Unless workers and the public have a clear understanding of the risks and how the risks are managed, they may be reluctant to buy into the technology and any particular risk-reduction strategy.

If a small group of individuals is asked to characterize a particular activity as risky or not (say, automobile travel and airline travel), a wide variety of responses is likely. Individuals interpret risk in different ways. People rate risks differently because, among other things, some individuals may know more about the risk than others. Knowledge includes technical familiarity with the activity or agent as well as personal experience. Groups that have different knowledge bases (e.g., lay persons vs. experts) often display dramatic differences in how risk is perceived. It should not be inferred from this that experts are always right because they are "experts." On the contrary, experts may be no better equipped to gauge the severity of some risks because they may discount important nontechnical consequences such socio-economic costs or cultural impacts of the technology.

Risk communication is important because public perceptions of risk do not always match the actual risks. People fear the wrong things. We fret about activities that involve small risks and do not pay enough attention to risks that are significant and about which we can do something. Consider automobile travel and airplane travel. Many people will not fly but have no hesitancy about getting into a car. In the 1990s Americans were, on a mile-for-mile basis, 37 times more likely to die in a car crash than on a commercial airliner. Commercial airline travel is so safe that the chances of dying in any flight are less than tossing heads 22 times in succession.[31] Although the risks are substantially higher for automobile travel, people do not seem to think the risks are anything to worry about. According to the National Highway Traffic Safety Administration, automobile traffic safety belts save about 9,500 lives per year. When used properly seat belts reduce fatal injury risk to front-seat car passengers by 45%. More than 25% of Americans do not use seat belts.

Risk communication is a two-way street. Experts need to communicate risk information in understandable ways. Use of single-number estimates is unjustified because they convey a sense of certainty that is not usually borne out by the scientific data. Expressing risk in the form of a central estimate and confidence interval or range is more meaningful. But one must be careful not to attach too much weight to the upper and lower bounds of the confidence interval. Risk comparisons are useful in putting risks in perspective, but choosing the right comparisons is important because people perceive risks differently. Using smoking risks in a comparison will have different meaning for the smoker and the nonsmoker. In general, risks are most understandable when comparisons involve related agents, different sources of exposure to the same agent, different agents to which humans might be exposed in similar ways, or different agents that produce similar effects.[32]

The public has a role in enriching the risk dialogue by expressing perspectives and preferences. Decision makers need to hear from the public about their views on what may be considered negligible risk or what levels of risk may be considered acceptable. In this regard there is no single "public." Risk acceptability is a moving target. A risk may be acceptable to one group but not to another because of differences in perceived benefits of the activity or product as well as economic considerations.

Do we worry about the right things? According to accident statistics and death rates, apparently we do not. Cigarette smokers who worry about radiation from mammograms or chest x-rays have perceptions of risk that are not congruent with what we actually know about these risks. There is no evidence that chest x-rays and mammograms kill anyone. However, cigarettes kill more than 400,000 people every year from cancer and heart disease. Certainly whether the risk is considered voluntary or controllable impacts how it is perceived. There is substantial literature on the subject of risk perception.[33] Prioritization of risks and how risks should be expressed for effective communication are central themes in this book. These issues are addressed more completely in Chapters 7 and 10.

The next chapter focuses on the dose-response function as the centerpiece of the risk assessment process. The shape of the curve and its slope determine how risky a particular agent is. The dose response is at once the strength and the weakness of the risk-assessment process. Dose response allows for quantifying risk at specific exposure levels and also predicting risk where epidemiological evidence may be absent. This information is vital to policy and public health decision makers. But dose-response assessment also has great uncertainties, particularly when dose extrapolation is large. Under great uncertainty risk estimates may be of little value to decision makers. For many environmental agents doses are so small that resulting risk estimates (based on application of dose-response theories) are highly uncertain and of little, or no, value in the decision-making process. This book argues that the role of the dose-response assessment in the risk-assessment process needs to be reconsidered. When considering very small risks, the dose-response assessment has little utility in the risk-assessment process. Risk must be assessed using other strategies such as comparing doses to natural levels (if the agent occurs naturally). Chapter 4 discusses problems of uncertainty and the difficulties encountered when large dose extrapolations are employed. Often scientific data are not clear enough, particularly in the low-dose range, to distinguish one dose-response function from others. When the science is unclear, decision makers resort to the simplest and most conservative theory to predict risk. Although this may be a safe approach, the most conservative theory may not necessarily be the right one.

NOTES AND REFERENCES

1. Prenatal development is a high-risk phase of life. Half of all pregnancies in the U.S. result in prenatal or postnatal death or an otherwise less than healthy baby. Early pregnancy loss (during first 8 weeks of pregnancy) occurs in 20% to 30% of implantations. Spontaneous abortions (8 to 20 weeks of pregnancy) occur in 10% to 20% of clinically recognized pregnancies. Minor developmental defects occur in about

15% of live births. Major congenital anomalies occur in 2% to 3% of live births. Chromosomal defects appear to be a major cause of spontaneous abortions and anatomical malformations. See National Research Council, Committee on Developmental Toxicology, *Scientific Frontiers in Developmental Toxicology and Risk Assessment,* National Academy Press, Washington, DC, 2000.

2. Most diseases, including cancer and heart disease, occur as a result of some combination of genetic and environmental risk factors. Some individuals are predisposed to certain diseases because of genetic factors that cannot be changed at this time. This explains in part why some never-smokers get lung cancer when over 90% of lung cancer occurs in smokers. Eliminating environmental factors such as smoking or alcohol consumption may not reduce the disease risk entirely because of residual risk associated with the individual's genetic make-up. In the future it may be possible to eliminate disease causing genetic mutations through gene therapy technology.
3. Brushing teeth can result in bleeding gums. For individuals with certain heart conditions (e.g., valve replacements), this can be a serious matter because oral bacteria can enter the bloodstream and cause heart damage.
4. Certain diagnostic x-ray procedures such as fluoroscopy may involve large doses resulting in burns and other acute effects. See National Cancer Institute, *Interventional Fluoroscopy: Reducing Radiation Risks for Patients and Staff,* NIH Publication No. 05-5286, March 2005.
5. Mossman, K.L., Analysis of risk in computerized tomography and other diagnostic radiology procedures, *Computerized Radiology,* 6, 251, 1982.
6. See Ropeik, D. and Gray, G., *Risk: A Practical Guide for Deciding What's Really Safe and What's Really Dangerous in the World around You,* Houghton Mifflin Company, New York, 2002. This catalog provides a description of a wide variety of everyday and technological risks. Included in their analysis is a breakdown of each risk by the probability of occurrence and severity of consequences.
7. Nuclear power plants overall have an impressive safety record. The Three Mile Island nuclear power plant accident in 1979 caused serious damage to the plant, but there were no deaths or injuries to workers or the general public as a result of releases of radioactive material to the environment. The Chernobyl accident in 1986, on the other hand, caused serious environmental and public health effects because of deliberate inactivation of safety systems. This resulted in massive releases of radioactive material to the environment from a reactor with minimal containment capabilities.
8. About 50 million adults smoke in the U.S. Men are more likely to smoke than women. The highest percentage of smokers is in the 18- to 24-year-old group. Of the approximately 400,000 deaths per year attributable to smoking, about 160,000 are from lung cancer. See American Cancer Society, *Cancer Facts and Figures 2005,* American Cancer Society, Inc., Atlanta, GA, 2005.
9. These assumptions are not entirely valid but nevertheless serve to illustrate how risk calculations are made. Individuals in the population are not at equal risk; accidents are not uniformly distributed in the population.
10. Approximately 85,000 Japanese survivors of the atomic bombs dropped on Hiroshima (August 6, 1945) and Nagasaki (August 9, 1945) that ended World War II have been continuously monitored for health effects that may have been caused by nuclear radiation. About 10,000 survivors received a sufficiently high radiation dose to increase cancer risks significantly.
11. The definition of low risk is based on a consideration of health effects from x-rays. More is known about the human health effects of radiation than any other chemical or physical agent except perhaps cigarette smoke (composed of thousands of chemicals).

Studies of the Japanese survivors of the atomic bombings and other epidemiological investigations suggest that doses below 0.1 sievert (a unit of radiation dose) are not associated with statistically significant risks of cancer. Using a lifetime cancer mortality risk estimate of 0.02 radiation-induced cancers per sievert (Sv) for chronic exposures, the lifetime risk is roughly 1:1,000.

12. Regulatory limits for exposure to carcinogens have been set at risk levels significantly below 1:1000. Under the Comprehensive Environmental Response, Compensation and Liability Act of 1980 (CERCLA), EPA regulates chemical carcinogens based on a lifetime cancer mortality risk of 1:10,000. CERCLA was enacted in response to the growing problem of improper handling and disposal of hazardous substances. CERCLA authorizes the EPA to clean up hazardous waste sites and to recover costs associated with the cleanup from entities specified in the statute. The superfund statute is the primary federal law dealing with the cleanup of hazardous substance contamination.
13. The Delaney Clause, introduced by Congressman James J. Delaney (New York), appears in the Food Additives Amendments to the Food, Drug, and Cosmetics Act of 1958. This legislation regulated pesticide residues as food additives. The 1996 Food Quality Protection Act effectively repealed the Delaney Clause by redefining pesticides as non-food additives. However, the Delaney Clause is still in force with respect to color additives.
14. Beck, M. et al., "The Tylenol Scare," *Newsweek*. October 11, 1982.
15. Lowrance, W.W., *Of Acceptable Risk: Science and the Determination of Safety,* William Kaufmann, Inc., Los Altos, CA, 1976.
16. National Research Council, Committee on the Institutional Means for Assessment of Risks to Public Health, *Risk Assessment in the Federal Government: Managing the Process,* National Academy Press, Washington, DC, 1983. This National Research Council report is often referred to as "The Red Book."
17. Ibid.
18. A.B. Hill identified several criteria for establishing causality: (1) the strength of the statistical association between agent and health outcome, (2) the presence of a dose-response relation, (3) consistency of results across studies, (4) temporal order (exposure precedes appearance of disease), (5) plausible biological mechanism, (6) coherence of evidence, (7) specificity (i.e., the disease is caused only by the subject agent), (8) analogy, and (9) experiment (e.g., reduction in dose of agent results in decrease in disease). Causality can be very difficult to establish because rarely are all criteria met. In the case of ionizing radiation, biological mechanisms for cancer induction are known, and consistency of results across studies has been remarkable. However, the demonstrated statistical associations have not been strong except in a few instances, and disease specificity has been largely absent. See Hill, A.B., The environment and diseases: Association or causation? *Proceedings of the Royal Society of Medicine,* 58, 295, 1965.
19. The thalidomide tragedy showed the limits of animal models as predictors of human response. In 1961, clinical evidence became available that a mild sedative, thalidomide, when taken early in pregnancy caused an enormous increase in a previously rare syndrome of congenital anomalies involving the limbs. Prior to withdrawal from the market in 1961, about 7,000 affected infants were born to women who had taken this drug. A single dose of the drug was sufficient to produce effects. The usual preclinical animal tests did not detect that thalidomide had teratogenic activity because different animal species metabolize the drug differently. Pregnant mice and rats (the most common test animals) do not generate malformed offspring when given thalidomide. Rabbits produce some malformed offspring, but the defects are different

from those seen in affected human infants. An excellent overview of the thalidomide tragedy is provided in The Insight Team of the Sunday Times of London, *Suffer The Children: The Story of Thalidomide,* The Viking Press, New York, 1979.

20. Environmental Protection Agency, *Updated Draft of Guidelines for Carcinogenic Risk Assessment,* Office of Research and Development, Environmental Protection Agency, Washington, 1999; International Agency for Research on Cancer (IARC), *Overall Evaluations of Carcinogenicity to Humans*, http://monographs.iarc.fr/monoeval/crthall.html (accessed March 2006).
21. U.S. Department of Health and Human Services, Public Health Service, National Toxicology Program, *Report on Carcinogens*, 9th ed., 2000.
22. U.S. Department of Health and Human Services, Public Health Service, National Toxicology Program, *Report on Carcinogens*, 11th ed., 2005.
23. Some radioactive substances can come into direct contact with the individual through deposition on the skin, inhalation, or ingestion. The behavior of the radioactive material after it is internalized is governed by its chemical properties rather than its radiological properties. For example, inhalation of radioactive iodine results in uptake of iodine into the thyroid gland where it is used to synthesize thyroid hormone. Radioactive iodine has essentially the same chemical properties as stable iodine and is handled by the body in the same way. This is a basic principle of nuclear medicine—a medical specialty that uses radioactive material in the diagnosis and treatment of disease. Depending on the amount of iodine taken up by the thyroid, radiation emanating from the radioactive atoms may damage the thyroid gland. In the treatment of thyroid cancer, some patients may be given radioactive iodine to destroy any remaining thyroid tissue after surgical removal of the gland.
24. Sexton, K., Needham, L.L., and Pirkle, J.L., Human biomonitoring of environmental chemicals, *American Scientist*, 92, 38, 2004.
25. *Supra* note 16.
26. Presidential/Congressional Commission on Risk Assessment and Risk Management, *Risk Assessment and Risk Management in Regulatory Decision-Making,* Final Report Volumes 1 and 2, 1997. http://www.riskworld.com/riskcommission/Default.html (accessed March 2006).
27. Ibid.
28. U.S. Nuclear Regulatory Commission, *White Paper on Risk-Informed and Performance-Based Regulation*, SECY-98-144, June 22, 1998, http://www.nrc.gov/reading-rm/doc-collections/commission/secys/1998/secy1998-144/1998-144scy.html#ATTACHMENT (accessed March 2006); Nuclear Regulatory Commission, Medical use of byproduct material; policy statement, revision, *Federal Register*, 65, 150, 47654, August 3, 2000.
29. Vetter, R.J., Medical health physics: A review, *Health Physics*, 88, 653, 2005.
30. Gigerenzer, G., *Calculated Risks: How to Know When Numbers Deceive You,* Simon & Schuster, New York, 2002.
31. Myers, D.G., Do we fear the right things? *Skeptic*, 10, 1, 56, 2003.
32. *Supra* note 30.
33. Slovic, P., *The Perception of Risk*, Earthscan Publications, Ltd., London, 2000.

2 Scientific Guesswork

Chapter 1 reviewed the elements of risk assessment, a process in which predictive theories are used to estimate health effects at low doses because direct observations are not usually available. In the current system of chemical and radiation protection, risk information at small doses is necessary in regulatory decision making and in risk-management practices. Decision makers and regulators need a handle on risks to understand the public health impacts of dose limits and risk-management options. Risk-management practitioners use risk information to develop and implement cost-effective risk-reduction strategies.

Choosing the right dose response has broad public health, political, social, and economic implications. If the dose-response function overestimates the risk, huge sums of money may be expended needlessly to reduce public health and environmental risks that either do not exist or are too small to be of concern. If the theory underestimates the true risk, avoidable harm to society, public health, and environment may result. If the dose response is steep (i.e., high risk coefficient), even a small reduction in dose can lead to a dramatic diminution in risk. A shallow dose response indicates that changes in dose are not likely to affect risk in a significant way. Risk-communication strategies also hinge on the shape of the dose-response curve. "Safety" is much easier to sell if the underlying dose response has a threshold. Below the threshold dose, risk of adverse effects is zero. The notion of "safety" is more challenging if the dose response is without a threshold because it can be interpreted that any dose is potentially harmful.

The costs of environmental remediation of federal sites used for the development of nuclear weapons during the Cold War illustrate the magnitude of the problem in an economic context. The U.S. government may spend as much as $350 billion to clean up sites in the nuclear weapons complex to comply with the U.S Environmental Protection Agency (EPA) cleanup standard of 0.15 mSv per year. Costs might be reduced by as much as $200 billion by relaxing the cleanup standard under the assumption that the less restrictive dose limit does not negatively impact public health or the environment.

This chapter focuses on the general types of dose-response theories that can be used in risk assessment and risk management. Several overarching themes will guide this discussion. The first theme addresses the notion of observational limitations at small doses typically encountered in environmental or occupational settings. Risks are very small and difficult to detect. Cancer has a high spontaneous incidence, introducing a significant signal-to-noise problem.[1] Compounding this problem is the fact that, as a general matter, cancers induced by carcinogens are clinically indistinguishable from cancers that arise from no identifiable cause. Second is the general question of risk uncertainty at small doses as a consequence of the need for large dose extrapolation and cross-species extrapolation. The notion of uncertainty is a theme that pervades this entire book. An appreciation of the sources and magnitude of uncertainty in risk estimates is important in operational risk assessment and management. Third is a corollary to the second theme and addresses the question of thresholds. Is any carcinogen dose potentially harmful, or must a certain nonzero dose be exceeded? The question of thresholds is important because it is a rational upper bound for establishing a regulatory dose limit, and may also serve as a philosophical basis for risk management that does not require reduction of dose to zero.

MAKING THE RIGHT CHOICE

What characterizes an appropriate dose-response theory in risk assessment and management? From a scientific perspective, the "right" theory should provide a reasonable fit to all of the scientific data currently available. Using selected data to support a particular theory to the exclusion of counterevidence is unacceptable scientific practice. The most appropriate theory is the best fit to all of the data. Social, economic, and political factors are important criteria in theory selection.

From a science perspective, selection of one theory to the exclusion of biologically plausible alternatives should be based solely on the scientific evidence. However, in current cancer risk assessment, the scientific evidence in humans at very low doses is inadequate to eliminate candidate theories in favor of a single theory. From a policy perspective, economic and other nonscientific factors also drive theory selection. Often the scientific and policy perspectives are not congruent and may be at cross purposes. Even if there is overwhelming scientific evidence, a particular theory may be inappropriate from a political, social, or economic perspective. For instance, a threshold defined theory may never be viewed as politically acceptable because of public concern that even very small doses may carry risk even though the available data suggest otherwise. Theory selection may boil down to factors that have little to do with science. The "right" theory may not necessarily be the one that is the best fit to the scientific data.[2]

Low-dose risk projections are not scientifically defensible because data are inadequate to select one theory to the exclusion of alternatives, and risks at low doses have large uncertainties that preclude utility in risk management decisions. A more scientifically defensible approach would be to avoid risk projections altogether (and the problem of choosing the "right" dose-response theory) and assess public health detriment by comparing doses from different sources. This approach, and its strengths and weaknesses, is discussed in detail in Chapter 7.

Simplicity and risk conservatism are key nonevidentiary determinants in dose-response selection for the purposes of risk assessment and risk management. Because of the paucity of scientific data at small doses and lack of human data for many agents (primarily chemical carcinogens), any biological plausible theory will be highly uncertain in its prediction of risk at environmental and occupationally relevant dose levels.

The first step in theory selection is fitting biologically plausible alternatives to the available scientific data. Theories should be excludable simply on the basis that they do not fit the data. Theory selection, in actuality, is much more complex than simply using a best fit-to-the-data criterion. The fitting process can be as simple as drawing by eye a straight line between data points or as complex as finding the best

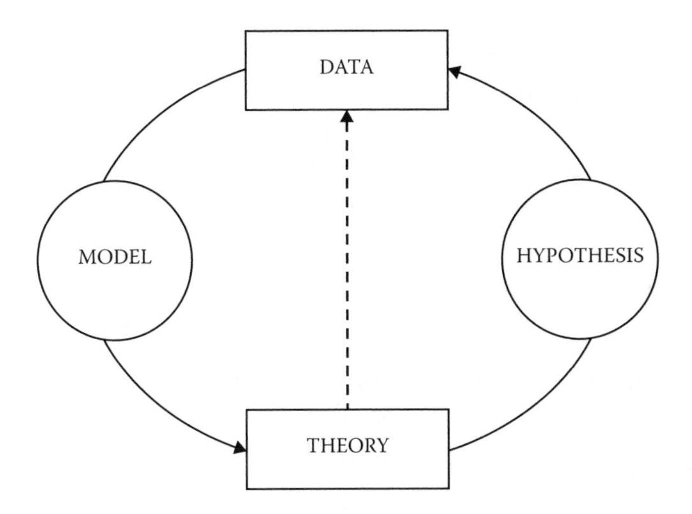

FIGURE 2.1 The dynamic relation between data and theory. The traditional approach to scientific inquiry is the process of hypothesis testing of a scientific theory. Experiments are conducted to obtain data as a test of a hypothesis to support or negate a particular theory. If the data do not confirm to theoretical predictions, the theory and its underlying model are either modified to account for the new observations or abandoned. Conceptual models are used to facilitate theory development. The dashed line indicates that the collection and interpretation of the data to test the theory are influenced by the theory itself.

statistical fit to the data by adjusting theory parameters using a computer. However, data fitting is seriously limited (particularly for human cancer) by the paucity of data in the dose range that distinguishes one theory from another.

As shown in Figure 2.1 theory selection is centered on an inductive approach (i.e., theories result from making observations and analyzing data) that requires an appreciation for the complex dynamic between data and theory. A scientific theory is a set of principles or statements to explain a collection of data, observations, or facts. When such statements or principles withstand repeated testing and attempts to falsify, they are established as a theory. Statements underlying a theory are often expressed in mathematical terms with biologically meaningful parameters. For example, Equation 2.1 describes the linear nothreshold (LNT) theory:

$$R_D - R_0 = \alpha D \qquad \text{(Equation 2.1)}$$

where R_D is the response (e.g., number of cancer deaths) following dose, D of the agent; R_0 is the response when $D = 0$ (e.g., spontaneous or natural incidence of cancer); and, α is the slope expressed as response (e.g., number of cancer deaths) per unit dose and is an expression of risk or biological sensitivity.[3] Actually the LNT equation above is a special case of a more general equation of the form

$$R_D - R_0 = \alpha D + \beta D^2 \qquad \text{(Equation 2.2)}$$

where βD^2 is a quadratic term that becomes important only at high doses.

This linear-quadratic equation is used by the International Commission on Radiological Protection (ICRP) and the National Council on Radiation Protection and Measurements (NCRP) in their recommendations on radiation risks to describe the relation between radiation dose and radiogenic health effects over a wide range of doses. At high doses cell killing predominates over cancer causation resulting in acute tissue injury. However, at environmental and occupational doses of interest in radiation protection, the dose-response relation can be described by LNT theory because the quadratic parameter (β) in Equation 2.2, a measure of the probability of cell killing, is essentially zero.

In the LNT formulation, risk per unit dose (i.e., the risk coefficient) is the same at any dose. Clearly, the assumption of dose independence has significant policy implications because the nature and mechanisms of effects at high dose are assumed to be identical at low dose. The shape of the dose-response curve provides some insight into the underlying biological processes leading to the observed responses. The LNT expression above suggests that a single molecular or cellular event is all that is needed to produce the effect. If the response is cancer, LNT theory suggests that tumors arise from a single aberrant cell.

A theory is a conceptual framework that facilitates organization of observations and other data in some meaningful way and also predicts future outcomes when theory-dependent conditions are specified. In the LNT example above, this would mean that α, D and R_0 are specified to predict the value of R_D. A good theory provides unexpected insights into observable phenomena. But theories are more than the result of analysis of patterns in the data. Theories involve novel concepts or ideas and suggest relationships that may not be evident from inspection of the data. Models (physical or mental constructs of the process or phenomenon) are a necessary tool in theory development.

Models connect data to theory and facilitate theory development by conceptualizing underlying processes or structures. For example, the Bohr model of the atom (which describes the movement of electrons around the dense atomic nucleus) was an important construct that led to the development of quantum theory in physics. Watson and Crick's proposed model for the structure of DNA in 1953 was critical to the development and advancement of molecular biology theory that has been central to the rapid progress observed in biotechnology and molecular medicine today.[4] As discussed in the next chapter, the target model of radiation action was an important step in the development of LNT theory.

Scientific theories become established through the accumulation of supportive scientific evidence. Evidence is accumulated through a process called hypothesis testing. Scientific inquiry has traditionally involved hypothesis testing in which a theory is tested through experimentation.[5] A hypothesis is a conjecture usually in the form of a question that is amenable to confirmation or refutation by experiment. Hypothesis is sometimes considered a synonym of "theory," but in the strictest sense it is a test of a theory. Hypotheses form the basis for the design of experiments to test a theory.[6]

The philosopher Karl Popper argues that theories can only be disproved and can never be proven absolutely because the set of all possible tests of a theory can never be completed. One would never know if the next test would invalidate the theory.[7] Theories are subject to constant tests of validity. If a set of experimental observations does not support a theory, the theory must either be abandoned or modified to account for the new data or observations. A theory becomes widely accepted in the scientific community when it consistently withstands falsification. This requires that the theory be testable in a way that it is subject to falsification. Theories that are not testable have limited utility in science. Confidence in a theory and its utility in various science and engineering applications derives from resistance to falsification. The issue of testability has been a fundamental problem in theory selection in risk assessment. All of the biologically plausible dose-response theories considered in this chapter are untestable in the low-dose range.[8] As discussed below, it is the data in the low-dose range that distinguish theories. Because theories cannot be falsified based on low-dose evidence, it is impossible to select one particular theory to the exclusion of alternatives.

Testing a theory is not a trivial matter. The data must be of sufficient rigor to withstand critical review.[9] Otherwise, data that are derived from faulty experimental designs or methods can simply be discounted in defense of the theory.[10] Testing a theory must involve all of the evidence; selection of certain pieces of evidence in support of a theory is an inappropriate test of the theory.

Complicating the problem is the fact that data are theory laden. Theory impacts collection and interpretation in at least two ways. First, interpretation of data may be influenced by the underlying theory being tested, as suggested by the upward, dashed arrow in Figure 2.1. Second, design of experiments and collection of data are driven by theory-derived hypotheses. Theories are lenses through which the scientist makes and interprets observations. Scientific inquiry is often thought of as a purely objective and dispassionate enterprise that involves a detached observer collecting scientific data. Nothing could be further from the truth. Science is a highly subjective process that involves collection and interpretation of data based on the scientist's personal experiences and philosophical approaches (including theory preferences). Many scientific controversies are a result of legitimate differences of opinion concerning interpretation of the same data set. Hypotheses are often constructed in order to test one theory without due consideration of other theories that the investigator may not consider to be plausible.

Theories suggest hypotheses to be tested (Figure 2.1). Depending on the theory, an important hypothesis may not be tested. Scientists and institutions may have a vested interest in a particular theory that colors the way scientific experiments are conducted and interpreted. Adopting a restricted or narrow scope of inquiry in this way may necessarily exclude valid alternative perspectives.

Theories predict a certain set of observations but not others. The accuracy of the predictions depends upon how well the theory fits observations.[11] The quantity and quality of the scientific data should be the basis for theory selection. By quantity, we mean that data are available over a wide range of doses of the agent. Statistically significant data at low doses are particularly important because that data distinguishes theories. By quality, we mean that the data are derived from experimental methods and designs that are rigorous and acceptable to the scientific community.

PREDICTIVE THEORIES IN RISK ASSESSMENT

Theories may be classified according to their risk predictions at low doses. Dose-response curves are used to predict risks at low doses when direct observations are available only at high doses. As shown in Figure 2.2, theories tend to converge and cannot be readily distinguished at high doses. But the theoretical risks are very different at small doses. Risks may be negative (i.e., the agent confers a beneficial effect) or zero (the dose response has a threshold), or they may be positive depending on the shape of the dose-response curve. As a consequence, choosing a particular theory has serious economic, political, and social impact.

Because of the paucity of significant data in the low-dose range, theoretical estimates of risk using any of the dose responses in Figure 2.2 have substantial uncertainty. The uncertainty is so great that the utility of theory-derived risk estimates in decision making must be called into serious question. In the end it really does not matter what predictive theory is used. Risk estimates at low doses have large enough uncertainties to include risk predictions by almost every biologically plausible dose-response theory.

LINEAR NO-THRESHOLD THEORY

LNT (Figure 2.2) is the preferred theory used by federal agencies such as the EPA and the U.S. Nuclear Regulatory Commission (U.S. NRC) to calculate cancer risks at doses encountered in occupational or environmental settings. The assumption is that risk extends to zero dose and that health effects are proportional to dose. There is no dose below which the risk is zero. This theory has emerged as the dominant predictive theory in risk assessment and risk management because it serves as a reasonable middle ground in risk assessment. Although no single theory satisfactorily

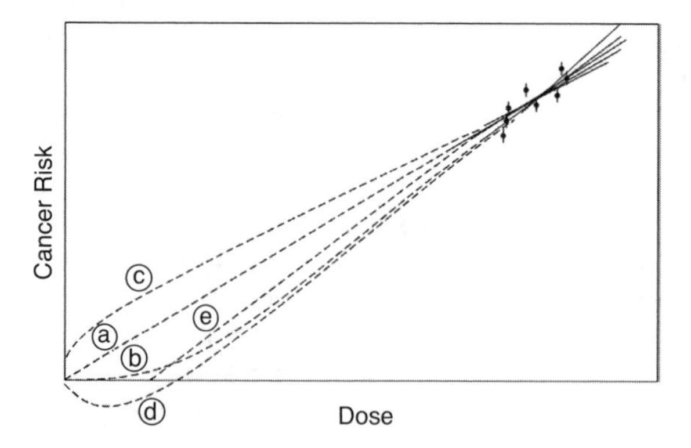

FIGURE 2.2 Possible shapes of dose-response curves in risk assessment: (a) linear, no threshold, (b) sublinear, (c) supralinear, (d) hormesis or U-shape, and (e) threshold. Dose-response curves fit the data equally well at high doses but predict very different risks at low doses. The data points (with error bars) and the solid lines represent the region of direct observations; the dotted lines represent theoretical risk projections.

fits all of the data, LNT provides the most consistent fit to the broadest range of experimental animal and human epidemiological data.

Over the range of doses where direct observations of health effects can be made, there is compelling evidence that the probability of health effects is linear with dose and that the risk per unit dose is independent of the size of the dose.[12]

LNT theory also predicts the absence of a dose threshold. Proving no threshold is difficult because the chance of observing health effects is very small and absence of effects at a given dose level does not mean that effects are absent. Without clear evidence for a threshold, LNT proponents argue that it is prudent to assume no threshold. Based on what is known about cancer induction and the difficulties in measuring health effects at small doses, LNT theory would appear to provide a reasonable approach to risk assessment and risk management.

Because of the central role LNT plays in risk assessment and risk management, Chapter 3 is devoted to a detailed discussion of LNT, the validity of theory assumptions, and how LNT became the predominant theory in risk analysis.

Sublinear Nonthreshold

For some types of cancers (e.g., leukemia), the dose-response curve appears to be sublinear without threshold (Figure 2.2). This dose-response function predicts that the risk at low doses would be less than proportional and lower than the risks predicted by LNT at low doses. Unlike LNT, the sublinear theory assumes that multiple independent events must occur in order to produce cancer. Distinguishing sublinearity from linearity at low doses is difficult because the probability of events is small, and the differences in risks predicted by each theory may be too small to be resolved with confidence. Theories may be resolved if the difference in slopes at low dose are statistically significant.

Sublinear dose responses may be used to estimate lower bounds on risk when no threshold dose is assumed. The National Research Council BEIR III Committee used linear-quadratic and quadratic predictive theories (types of sublinear nonthreshold theories) to estimate the lower limit of cancer mortality risk from exposure to low-level ionizing radiation.[13]

The data shown in Figure 2.3 are consistent with a sublinear dose response without a threshold. However, at doses below about 35 rad, the data points are also consistent with a threshold since the leukemia incidence cannot be statistically distinguished from the spontaneous or zero dose incidence. At doses 50 rad and higher, leukemia incidence is greater than the spontaneous incidence.[14] This figure clearly illustrates the difficulty in defining the shape of the dose response at low dose. The possibility of a threshold cannot be ruled out by the data because the probability of inducing leukemia is small. Testing for a threshold is possible if a sufficiently large number of test animals was available at several low-dose points. However, this may not be a practical approach because of the costs involved in conducting such an experiment. The ongoing study of the Japanese survivors of the atomic bombs provides a substantial database involving tens of thousands of individuals exposed to very small doses of ionizing radiation. Even this database is not sufficient to completely rule out a threshold.[15]

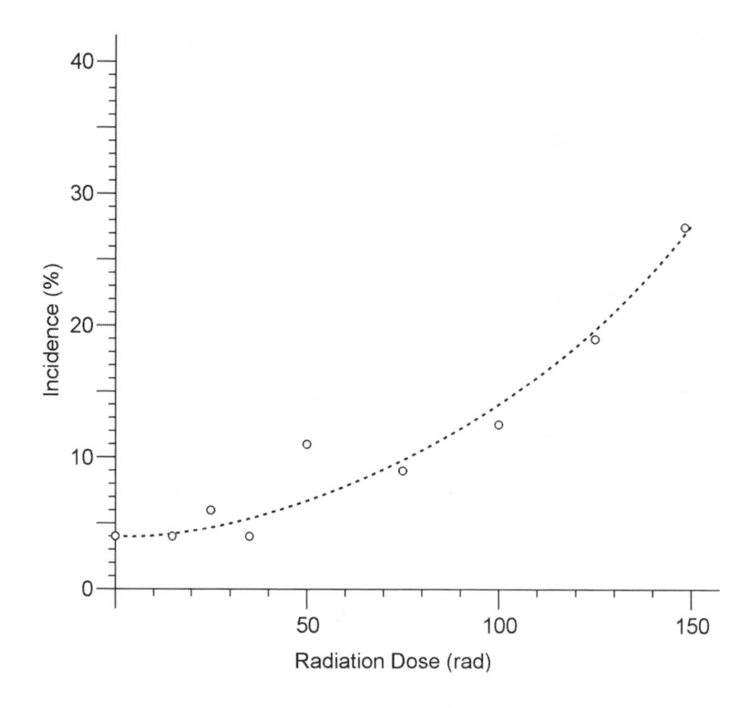

FIGURE 2.3 Sublinear dose response. The incidence of leukemia in mice exposed to whole body x-irradiation follows a sublinear dose response. Dose is expressed in "rad," the unit of absorbed dose of ionizing radiation. (Modified from Upton, A.C., The dose response relation in radiation induced cancer, *Cancer Research,* 21, 717, 1961, with permission.)

Supralinear

Supralinearity (Figure 2.2) argues that small doses are more dangerous than linear theory predictions. Supralinear dose responses project that risks per unit dose are higher at lower doses than at higher doses. Mechanisms to explain a supralinear dose response are complex. They likely include kinetics of cell and tissue damage and repair. For chemical agents, metabolism and detoxification are also important. Although there is evidence for supralinearity for some carcinogens, this dose response is not frequently encountered.[16] Minor changes in dose under supralinearity would have a greater risk reduction (i.e., health benefit) outcome than predicted by linear or sublinear theories. Figure 2.4 is an example of a clear case for supralinearity. This study was conducted in laboratory rats, and it is unclear whether the human response is similar.

Hormesis

There is a growing school of thought that small doses of agents may actually be beneficial and not harmful (Figure 2.2).[17] It has been known for centuries that small amounts of some substances may have completely opposite effects compared to large doses.[18] Alcohol, caffeine, and nicotine are poisons (inhibitors) at large doses

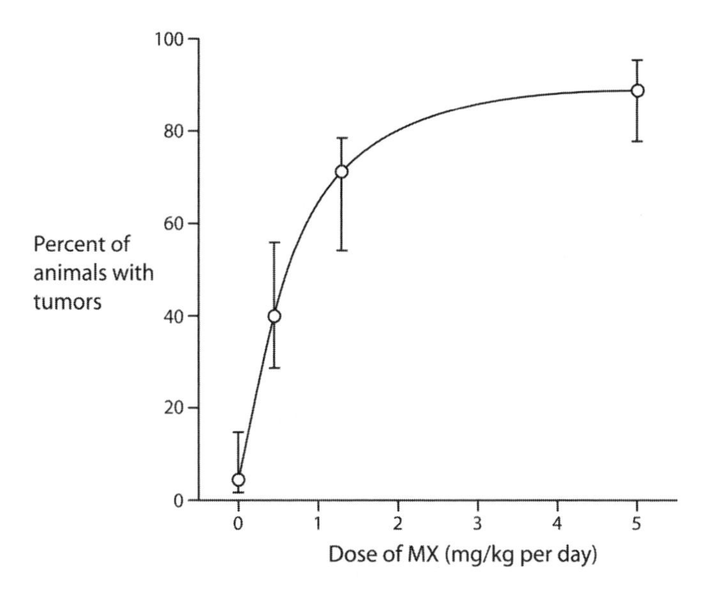

FIGURE 2.4 Supralinear dose response. Cancer induction in rats given the mutagen 3-chloro-4-(dichloromethyl)-5-hydroxy-2(*5H*)-furanone in their drinking water. (From Komulainen H. et al., Carcinogenicity of the drinking water mutagen 3-chloro-4-(dichloro-methyl)-5-hydroxy-2 (*5H*)-furanone in the rat, *Journal of the National Cancer Institute,* 89, 848, 1997.)

but in small doses act as stimulants. Even water exhibits dual properties! Water in small quantities is essential for life, but at very high doses death can result from excess hydration. This paradoxical effect is known as "hormesis" and generates a U-shaped dose-response curve (Figure 2.5). This dose-response curve is far more complex than others because of its biphasic nature.

Unlike LNT, hormesis posits that competing biological processes are at work and predominance of one process over another is dose dependent. At small doses, beneficial processes predominate (as reflected by a negative, downward slope in the dose response), but at high doses the opposite is true with a shift in slope direction from negative to positive.

Hormesis may be viewed as a special case of a threshold dose-response function in which subthreshold doses confer a beneficial effect (i.e., negative risk). Hormesis has not been considered favorably by standards-setting agencies or by national and international scientific advisory bodies.[19] There is little evidence in support of hormesis (as defined by reduced cancer mortality) in human populations exposed to small doses of radiation.[20] Hormesis is a complex biological phenomenon that remains poorly understood. The saccharin example in Figure 2.5 shows that hormetic effects may be gender dependent. The applicability of hormesis in risk assessment and management will remain questionable until gender and other significant biological determinants of hormesis are identified and understood.[21]

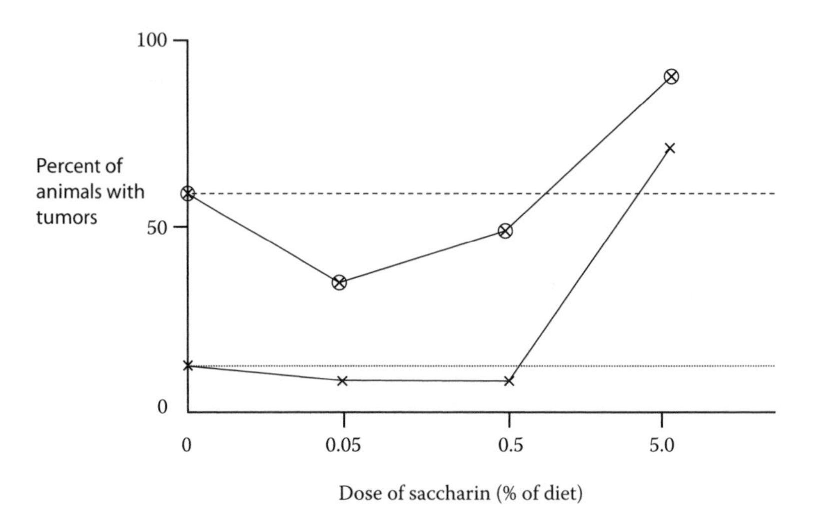

FIGURE 2.5 Hormesis dose response. Bladder tumors in female (upper curve) and male (lower curve) rats following saccharin exposure. The horizontal line through the upper curve is spontaneous incidence of tumors in female rats. The horizontal line through the lower curve is the spontaneous incidence of tumors in male rats. The locus of the threshold dose (zero effect point) is the intersection of the dose-response curve with the respective horizontal line representing spontaneous incidence. (From Office of Technology Assessment, *Cancer Testing Technology and Saccharin*, U.S. Government Printing Office, Washington, DC, 1977.)

THRESHOLD

A threshold theory (Figure 2.2) posits that a certain amount of damage must be accumulated before health effects are observed. This suggests that a minimum dose of the cancer-causing agent must be accumulated. The threshold dose response itself provides little information on the biological basis for the threshold. Molecular and cellular mechanistic studies and pharmacological studies (in the case of chemical agents) are needed to determine the importance of agent detoxification or neutralization, damage repair, and other biological processes.

Threshold detection is problematic and depends on how the study is conducted and the experimental methods used. It may not be possible to distinguish reliably between real zero effects (a true threshold) and effects that occur with such low probability that they cannot be measured reliably (an apparent threshold). The problem of distinguishing very low probability events from zero is discussed in Chapter 4. Furthermore, thresholds may be difficult to observe because of how the data are analyzed and plotted. Figure 2.6 shows the dose-response relationship for bone cancer in female radium dial painters. Radium is radioactive and emits energetic alpha particles (helium nuclei). When radium is ingested, it behaves like calcium and accumulates in bone. Essentially the same data were used to construct the different dose-response curves shown in Figure 2.6. The shapes differ because of how the data are analyzed and plotted. In each case bone dose is plotted on the *x*-axis, but the bone cancer metric on the *y*-axis is different. The Mays and

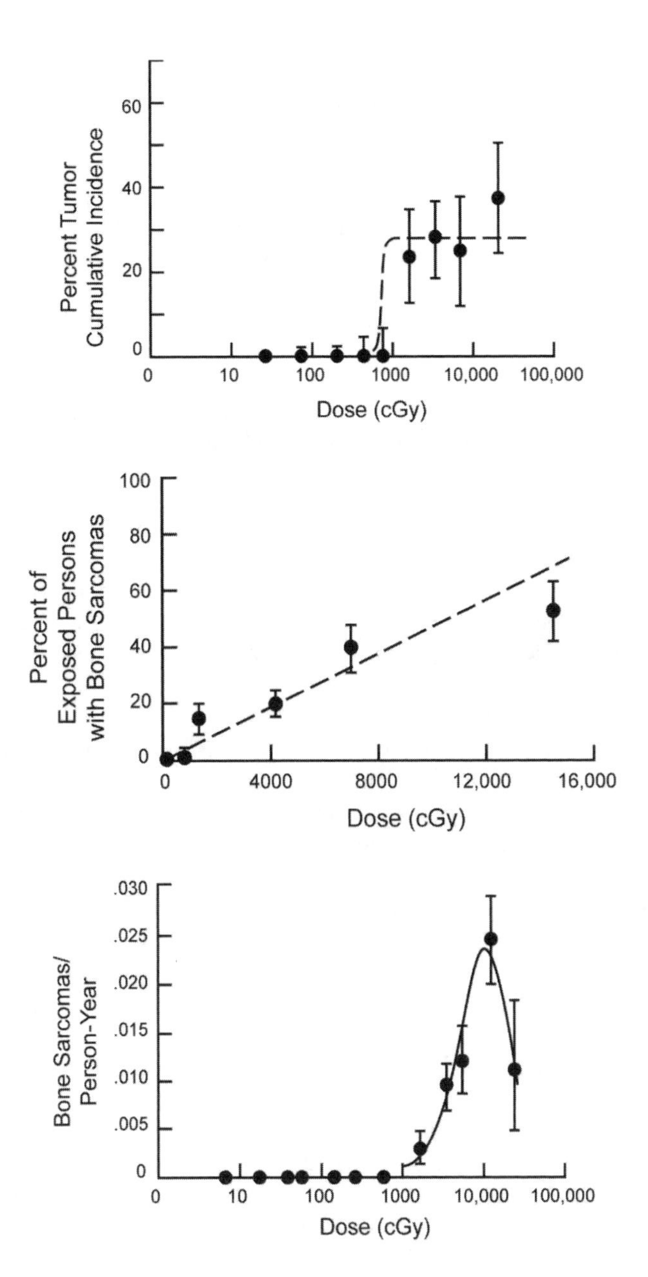

FIGURE 2.6 Thresheld dose response. Bone cancer in radium dial painters exhibits a dose threshold. Dose-response curves for three different analyses of bone cancer data in female radium dial workers exposed before 1930 are shown. The appearance of a threshold dose depends on how the data are plotted. (Modified from National Research Council, *Health Effects of Radon and Other Internally-Deposited Alpha-Emitters,* BEIR IV Report, National Academy Press, Washington, DC, 1988.)

Lloyd plot leaves a strong impression that the dose response is linear without a threshold but that is because the dose axis is significantly compressed. When the dose axis is expanded as shown in the other two studies, the threshold is readily apparent.[22]

The existence of a threshold provides a natural benchmark for the establishment of dose limits and risk-management goals. The problem is confirming the existence of the threshold in a variety of exposure situations. The size of the threshold is not likely to be constant in the population at risk. Sensitive subgroups likely have a smaller threshold or no threshold compared to others in the general population.

LIMITATIONS AND UNCERTAINTIES

The theories discussed above project very different risk estimates at low doses. The upper estimate is determined by the supralinear dose-response theory; the lower estimate is determined by the hormesis or threshold dose response. The range could be very broad (an order of magnitude or more) depending on the dose and the initial slopes of the dose responses curves. LNT theory projects numbers of health effects that are in the middle of the range and is preferred because it represents a conservative, central estimate.

There are considerable uncertainties regardless of which predictive theory is used. Cross-species extrapolation and dose extrapolation are perhaps the most significant sources of uncertainty in risk assessment.[23] Cancer risks from chemical agents are primarily based on animal tests since little, if any, useful human data are available. Rodents are typically used to evaluate carcinogenic effects in lieu of human experience. Cross-species extrapolation introduces substantial uncertainties in the form of overly conservative assumptions about human risk if it is assumed that cancer in test animals translates into cancer in humans at any dose. Laboratory animals are used on the supposition that chemical agent uptake, metabolism, storage, and excretion are similar to that in humans, and that if the agent is carcinogenic in the test animal it will also be carcinogenic in humans.

For ionizing radiation cross-species extrapolation is not an issue because human data are available to estimate cancer risk directly. Human cancer risks are based primarily on the study of the Japanese survivors of the atomic bombs. This is one of the largest studies of the effects of radiation on human health. The study (a joint effort between the U.S. and Japan) has been ongoing for over 50 years and involves more than 85,000 individuals for whom detailed medical records and radiation doses are available. The Japanese survivor study also involved a wide range of radiation doses that facilitated definition of the dose-response curve. Although doses approximating environmental and occupational exposures are not associated with discernible health effects, larger doses demonstrated clear cancer-causing effects. There is also significant data from the observation of cancers in patients given ionizing radiation for medical diagnostic or therapeutic purposes. Results of animal studies also support the direct estimates of cancer risks in humans. The congruence of human and animal carcinogenic responses is not surprising because physicochemical, biochemical, cell, and tissue responses following ionizing radiation exposure are expected to be species independent. For chemical agents, this assumption is not tenable because chemical agents may be absorbed and metabolized differently.

Dose extrapolation is also a serious source of uncertainty. For most carcinogens (including ionizing radiation), very large doses of the agent are needed to observe a statistically significant increase in cancer. Small doses typically encountered in environmental and occupational settings are associated with very low risks of cancer, and in the absence of any exposure, cancer occurs at a very high rate naturally (about 1 in 3 Americans will get cancer). Animal tests to predict human cancer risks are more likely to overestimate human risk because, in most instances, the carcinogenic response observed in test animals occurs at doses in excess of those anticipated to be encountered by humans.

Table 2.1 illustrates the magnitude of the dose extrapolation problem. Predicting radiogenic health effects at environmental and occupational exposure levels requires that directly observable dose response data be extrapolated two to three orders of magnitude (i.e., 100 to 1,000 times lower). This degree of dose extrapolation strains the credibility of risk assessment at low dose and is comparable to the dose extrapolations used to "demonstrate" the human cancer-causing effects of commonly occurring chemicals, including cyclamates, saccharin, and Alar, based on studies in experimental animals. Accordingly, numbers of cancer deaths due to low doses of carcinogens must be considered speculative; risk estimates at low dose have great uncertainties because they are theoretically derived. For ionizing radiation, the possibility that there may be no health risks from doses comparable to natural background radiation levels cannot be ruled out; at low doses and dose rates, the lower limit of the range of statistical uncertainty includes zero.[24]

TABLE 2.1
Extrapolating Health Risks

Agent	Equivalent Human Dose[a]	Typical Human Dose	Ratio[b]
Cyclamates	~500 cans of soda daily	1–2 cans of soda daily	~500
Saccharin	~1200 cans of soda daily	1–2 cans of soda daily	~1000
Alar	200,000 apples daily	0–1 apples daily	~200,000
Radiation from radioactive potassium	~3000 bananas daily	0–1 bananas daily	~3000

Sources: Cyclamate and saccharin data are from H.F Kraybill, Conceptual approaches to the assessment of nonoccupational environmental cancer. In *Environmental Cancer*, in Advances in Modern Toxicology, Volume 3, H.F Kraybill, H.F. and M.A. Mehlman, M.A., Eds., Hemisphere Publishing Corporation, Washington, DC, 1977. Data for Alar are taken from Fumento, M., *Science Under Siege,* Quill, New York, 1993. Ionizing radiation data are from Mossman, K.L., The linear no threshold debate: Where do we go from here? *Medical Physics,* 25, 279, 1998.

Notes

[a] Calculated human doses of carcinogen equivalent to actual doses used in animal experiments to demonstrate carcinogenic activity

[b] Ratio of equivalent human dose to typical human dose

It is difficult to justify labeling cyclamate, saccharin, and Alar as human carcinogens when doses necessary to produce cancers are orders of magnitude higher than doses typically consumed. As shown in Table 2.1, doses of agents to demonstrate carcinogenicity are so high that they are meaningless in the context of everyday uses.

Consuming hundreds of cans of soda daily is not only unrealistic, but individuals are likely to suffer the effects of overhydration long before any cancer might appear. Is it appropriate to label an agent as a human carcinogen when evidence of carcinogenicity can only be demonstrated at unrealistic doses?

SPECULATION VERSUS REALITY

Using dose-response theories to calculate health effects of exposure to very small doses of carcinogens is now so ingrained that real risks are no longer distinguishable from calculated, theoretical risks. Body bags are viewed the same whether they are real or calculated. Failure to distinguish reality from speculation poses enormous problems in risk assessment and management. Soon after the Chernobyl accident in 1986, various reports predicted thousands of deaths due to radiation exposure. In reality, the public health impacts have been far less.

The idea that no dose is safe, and concerns for "trivial risks," has contributed to a system of increasingly restrictive regulations. A 1997 ruling by the U.S. NRC illustrates the seriousness of the problem. In its final ruling on radiological criteria for license termination, the Commission determined that a site would be considered suitable for unrestricted release if the residual radioactivity, distinguishable from background, would result in an annual radiation dose above natural background that does not exceed 0.25 mSv.[25] This dose is about three orders of magnitude below doses associated with statistically significant radiogenic health risks in adult populations (as compared with background cancer rates), and is within the statistical variations of the natural radiation background in the U.S.[26] The EPA has an even more restrictive cleanup standard of 0.15 mSv per year in excess of natural background.

The U.S. NRC and EPA cleanup standards are set at a small fraction of the natural radiation background. Increases in cancer have not been observed at these limits because the probability of radiogenic effects is very low and the natural incidence of cancer in the population is high. Measurement of a radiogenic effect is confounded further by genetic variations in the population that are much greater than any radiation effect.

The idea that any dose is potentially harmful has led to unwarranted fears about radiation. In one survey of primary care physicians in Pennsylvania, 59% of the doctors identified fear of radiation as the primary reason for their patients' refusal of mammography examinations. Women who refuse mammography may be denying themselves an important medical benefit by compromising early detection and the subsequent management of disease. Following the Chernobyl accident, the International Atomic Energy Agency estimated 100,000 to 200,000 Chernobyl-related induced abortions in Western Europe. In Greece, as in other parts of Europe, many obstetricians initially thought it prudent to interrupt otherwise wanted pregnancies or were unable to resist requests from worried pregnant women in spite of the fact that doses were much lower than necessary to produce *in utero* effects.[27]

RISK MANAGEMENT AND RISK COMMUNICATION

The shape of the dose-response curve influences how risks are managed and communicated. Risk-management approaches are dictated by the magnitude of the initial risk and the expected costs and benefits of alternative risk-reduction strategies. Monotonic dose-response functions, as exemplified by LNT theory, are characterized by increasing risk with increasing dose. Accordingly, risk-management strategies focus on reducing dose to minimize risk. Ideally, dose should be reduced to zero to achieve zero risk (i.e., absolute safety), but this is practically impossible and quite costly. The more prudent course is to balance costs of dose reduction to achieve some level of benefit (i.e., risk reduction). The radiation-protection community uses the as low as reasonably achievable (ALARA) philosophy as its major risk management strategy. The central point of the ALARA approach is the requirement to balance costs of dose reduction as constrained by social and economic factors with expected public health benefits.

If the dose-response function is without a threshold, any dose carries some risk and decision makers must decide whether the risk is of sufficient concern to warrant action. If the dose-response curve is characterized by a threshold, then subthreshold doses have no associated health risks and further reduction is not necessary. An important practical problem is deciding the appropriate threshold level. There is likely a distribution of thresholds in the population since human responses to carcinogens are not homogeneous. The appropriate threshold should account for the most sensitive subpopulations. Threshold doses can be modified by the introduction of safety factors to account for heterogeneity in individual responses.

Interestingly, the threshold question may be moot. There is no reason to change regulatory limits because a threshold is identified below the dose limit. The difference between the limit and the threshold dose can be viewed as a safety factor. If the threshold is higher than established regulatory limits, it is unlikely that the dose limits will be made less restrictive by equating it to the threshold. Even in the face of compelling supportive scientific evidence, dose limits for human carcinogens are rarely relaxed.[28] In *Chlorine Chemistry Council v. EPA*, the court did not require revision of the EPA's no-threshold-based maximum contaminant level (MCL) for chloroform in drinking water (the enforceable legal standard) even though the agency had accepted new evidence of a threshold.[29]

Hormesis is a nonmonotonic dose response characterized by a biphasic or U-shape. There has been little discussion of how established risk-management strategies should be modified to account for the biphasic nature of hormesis. The simplest approach is to treat hormesis as a threshold dose response in which nonzero risks below the threshold are equated to zero. In this case risk management focuses on identifying the threshold level, including consideration of threshold heterogeneity in the population. However, as discussed earlier a threshold is not likely to impact standard setting.

If subthreshold beneficial effects are valued, dose-reduction strategies should avoid elimination of beneficial effects of the agent. Accordingly, the hormesis dose response needs to be fully characterized to identify the dose interval conferring beneficial effects. As suggested by Figure 2.5, this dose interval is subject to great uncertainty because of biological and agent variabilities, including gender differences and individual and subpopulation sensitivities. Under hormesis, increasing dose may actually be beneficial. But at what point should dose reduction be truncated to take advantage of any health benefits? Further, if the risk-management objective under hormesis is to maintain doses below a threshold (the dose above which detrimental effects occur), how can individual doses be managed when exposures to multiple sources (e.g., medical x-ray, occupational exposures) are possible? The usual risk-management objective of driving dose toward zero appears counterintuitive because small doses may be beneficial. The challenge to risk managers is developing strategies that reduce dose but only to levels that confer benefits. How such a program could be implemented and controlled is unclear and may be an intractable problem because of the potential for exposure to multiple sources.

Risk communication is an integral part of the risk-management process. How risks are communicated depends somewhat on the shape of the dose-response curve. For monotonic dose-response curves without a threshold, risk communications should emphasize that reducing doses to zero is not a practical goal. The notion that there is no safe dose and that any dose, no matter how small, is potentially harmful should be defused. Instead, the idea of acceptable dose needs to be promoted.[30] If the dose response is characterized by a threshold, risk communications may be less problematic since the threshold dose establishes a scientific "safety" point. A dose limit above the threshold is justified by the additional safety factor conferred by the dose savings (the difference between the limit and threshold). A dose limit below the threshold is clearly justified because it is set within the dose range associated with no health effects. Communicating risk under hormesis is perhaps the most challenging of all. Messages can be very confusing if, as some radiation hormesis proponents argue, "a little radiation is good for you." Under what circumstances is a dose "good" or "bad"? Is it the magnitude of the dose or the circumstances of exposure that is important? How do risk communicators clarify the mixed message that, on one hand, small doses of radiation may be beneficial outright (hormesis) and, on the other hand, small doses may be (theoretically) harmful but confer benefit through beneficial applications such as diagnostic medicine?

Regardless of the underlying dose response, greater emphasis needs to be placed on communicating the quality of the data used in the risk-assessment process. Estimates of risk derived only from laboratory animal data or cell studies should be so indicated to put the human risk in proper context. Confidence in data is linked to the quality of long-term animal studies, mechanistic studies, human epidemiologic studies and dose-response data. Uncertainties in cross-species extrapolation and dose extrapolation should be clearly articulated. The long-held belief that saccharin and other food additives are human carcinogens developed because the uncertainties associated with species and dose extrapolation were not fully appreciated or communicated. The more information conveyed, the higher the degree of public confidence. Decision makers in particular need to understand the sources and magnitude of risk uncertainties.

How numbers are used to express risk also impacts risk communications.[31] Using single numbers to express risk may reflect a level of confidence that is not supported by the available science. A more appropriate presentation would be a range of values with a central estimate. Given a range of values, policy makers usually adopt the upper limit to be as conservative as possible in decision making. However, this may be an unreasonable strategy, particularly if the range of uncertainty is large.

QUANTIFYING RISK AT SMALL DOSES

Risk assessors, risk managers, and policy makers use low-dose risk estimates to determine the number of people affected from an event that has already occurred or in "what-if" scenarios to calculate health detriment outcomes when certain population and exposure characteristics are assumed. Knowledge of health consequences provides guidance to decision makers about possible revisions in regulatory dose limits and in implementing risk-management scenarios. Using cost-benefit analysis and other quantitative and qualitative analytical methods, risk-management strategies can be evaluated if the health detriment is known within reasonable bounds of uncertainty.

The population health detriment (e.g., number of cancer deaths in the population due to agent exposure) is the product of the average individual dose of the agent, the size of the population at risk, and the risk coefficient (e.g., probability of getting cancer per unit dose of the agent). This last factor is heavily dependent on the choice of dose response. The confidence in the population risk is based on the uncertainty in the risk calculations. Each factor in the calculation has an uncertainty associated with it. Multiplying the factors to determine the total population detriment includes a quadratic propagation of the uncertainties associated with each factor. The total uncertainty in the health detriment is more than the sum of the uncertainties associated with individual factors.

Highly uncertain risks preclude sound decision making. Health effects from ionizing radiation exposure are better known than for almost every other human carcinogen, yet the uncertainty at low dose is substantial.[32] The situation for most chemical carcinogens is worse because most of the information used to estimate risks from chemical carcinogens derives from laboratory animal studies. The distinction between a real zero risk and a risk of zero because of methodological limitations is explored in Chapter 4. In the face of substantial statistical uncertainty, risk calculations at doses of occupational or environmental concern have little meaning since estimates can range from zero up to values determined by the dose-response specified upper bound of the confidence interval. The upper bound of the risk will be higher if LNT is selected than if a sublinear dose response is selected.

It is difficult to support the use of quantitative risk assessment for decision making when large dose extrapolations are required, and when the range of risk uncertainty spans several orders of magnitude (say from 1 in 10,000 to 1 in 1 million). Lower limits of 1 in 1 million are so small that they defy reliable measurement by current epidemiological methods. Accordingly, it is not unreasonable to establish a lower bound of uncertainty of zero. To the decision maker, such broad uncertainties translate into "anything is possible." How is the regulator or policy maker to use such information? Using the upper limit as a conservative estimate is unreasonable, particularly if there is epidemiological evidence of no health detriment. The upper bound of the 90% confidence interval is likely to seriously overestimate risk.

Quantitative risk estimation, although desirable, appears to have little utility in decision making when risks are very small. Methods that do not involve estimation of risk such as comparison of occupational or environmental doses with natural levels of the agent would seem a more reasonable and scientifically defensible approach to assessing health detriments. Unlike risk, exposures to most agents encountered in the workplace or environment can be measured directly.

The National Academies' BEIR VII Committee carefully reviewed the scientific and epidemiologic evidence on the health risks from low doses of ionizing radiation and concluded that LNT was the preferred theory. Other theories might provide a better fit to some of the data, but in the aggregate, LNT was judged to be most satisfactory. The BEIR Committee was careful to conclude that the available data could not exclude any biologically plausible theory outright. The Committee consensus in support of LNT was based on a preponderance of the evidence. Interestingly, the French Academy of Sciences reviewed the same epidemiological and radiobiological data and came to a different conclusion. In a report issued several months prior to the BEIR VII Committee report, the French Academy of Sciences concluded that LNT theory is not supportable by current scientific evidence.[33]

Whether LNT theory is appropriate for regulatory purposes at very low levels of exposure or for estimating cancer risks remains debatable. In searching for alternatives it is important to separate the issue of hormesis from thresholds. Convincing regulators and policy makers that thresholds exist and should be recognized in the standard setting process is a long-shot proposition. But there is an even lower chance that hormesis will be adopted because of the inconsistency of the hormesis evidence and how beneficial effects should be incorporated into standards. Introducing hormesis into the LNT debate almost guarantees that both hormesis and thresholds will be dismissed.

NOTES AND REFERENCES

1. Cancer is now the leading cause of death in the U.S. Roughly one in three individuals will get cancer; about one in five persons will die of cancer. See American Cancer Society, Cancer Facts and Figures 2005, America Cancer Society, Atlanta, GA, 2005.
2. Standards-setting organizations such as the U.S. Nuclear Regulatory Commission (U.S. NRC) and the Environmental Protection Agency (EPA) rely on independent authoritative bodies to provide analyses and evaluations of scientific evidence in support of their standard setting policies. In a 2001 report, the National Council on Radiation Protection and Measurements (NCRP) concluded that LNT theory is the most appropriate theory for radiological risk assessment because the available scientific data did not overwhelmingly support an alternative theory. The National Academies in its 2005 BEIR VII report drew similar conclusions. In 2005, the International Commission on Radiological Protection (ICRP) also endorsed the LNT theory noting that although the possibility of a threshold dose cannot be entirely eliminated the collective epidemiologic and scientific evidence does not favor the existence of a threshold. EPA and U.S. NRC continue to use LNT in risk assessment in accordance with the position that other theories are not more plausible. See National Council on Radiation Protection and Measurements, *Evaluation of the Linear-Nonthreshold Dose-Response Model for Ionizing Radiation*. NCRP Report No. 136, Bethesda, MD, NCRP, 2001; National Research Council, *Health Risks from Exposure to Low Levels*

of Ionizing Radiation, BEIR VII Report, National Academies Press, Washington, 2005; International Commission on Radiological Protection, Low-dose Extrapolation of Radiation-related Cancer Risk, ICRP publication 99, *Annals of the ICRP* 35(4), 2005.

3. In this formulation, R_D-R_0 represents the response due to the agent only. For ionizing radiation, R_0 cannot be determined directly because of irreducible natural background radiation. The percentage of spontaneous cancers due to background radiation has not been independently determined.
4. Watson, J.E. and Crick, F.H.C., A structure for deoxyribose nucleic acid, *Nature,* 171, 737, 1953. See also the January 23, 2003, issue of *Nature* containing a series of articles on the historical, scientific, and cultural impacts of DNA in celebration of the 50th anniversary of the discovery of the structure of DNA.
5. Hypothesis testing is one of numerous ways of doing science. The human genome project involved brute-force technology to identify the sequence of bases in human DNA. Developing and perfecting scientific methods and techniques are also legitimate forms of scientific inquiry that do not involve hypothesis testing. Discovery of the polymerase chain reaction (PCR) was critical to the astonishing progress in biotechnology. PCR, sometimes referred to as molecular photocopying, is a laboratory technique to make billions of copies of specific fragments of DNA. In criminal investigations, for example, PCR is used to make sufficient copies of DNA from tiny samples of biological material to carry out forensic analysis. Kary Mullis was awarded the Nobel Prize in 1993 for the discovery of the revolutionary technology.
6. Hypothesis testing does not necessarily have to be linked to theory confirmation. It may be used to establish a causal relation without inferring an underlying theory. For instance, in experimental psychology imaging different areas of the brain may assist in identifying causes of certain neurological diseases. Such studies are not conducted to support or refute a particular scientific theory but to identify proximal causation.
7. Popper, K.R., *The Logic of Scientific Discovery,* Basic Books, New York, 1959. The requirement that disagreement with data conclusively falsifies a theory is probably not correct. For instance, adding a correction factor to the theory that does not change its basic form or characteristics may be all that is needed to make the theory account for new data.
8. Dose-response theories in risk assessment are distinguishable by their shape in the low-dose region. However, little if any statistically significant data are available in the low-dose range to differentiate theories. Low-dose data are very difficult to obtain because the occurrence of health effects are low probability events and populations at risk are not sufficiently large to detect a health effect if it were present. Furthermore, health effects from agent exposure cannot be distinguished for spontaneous health effects in the general population. Identifying effects at low dose will require development of biomarkers of disease that are agent specific.
9. Data are not treated equally. Data that provide evidence for theory falsification tends to be scrutinized more carefully than supportive data. The expectation is that data will support established theory. Data are viewed with skepticism when it doesn't. Further, there is the argument that predictions (i.e., the hypothesis is formulated prior to collection and analysis of data) should count for more than accommodations (i.e., the hypothesis is constructed after the observations have been made). See Lipton, P., Testing hypotheses: prediction and prejudice, *Science,* 307, 219, January 14, 2005.
10. The recent controversy over a test of the linear no-threshold (LNT) theory using radon-induced lung cancer data is a good example. Cohen argues that his ecological

lung cancer data shows that LNT fails in the low-dose range. However, the quality of the data as a valid test of LNT has been called into question because the ecological experimental design was not well controlled. See Puskin, J.S., Smoking as a confounder in ecologic correlations of cancer mortality rates with average county radon levels, *Health Physics,* 84, 526, 2003; Cohen, B.L., The Puskin observation on smoking as a confounder in ecologic correlations of cancer mortality rates with average county radon levels, *Health Physics,* 86, 203, 2004; Puskin, J.S., James, A.C., and Nelson, N.S., Response to Cohen. *Health Physics,* 86, 204, 2004.

11. The National Research Council BEIR III report used LNT theory, linear quadratic theory, and quadratic theory to assess low-level radiation risks. Cancer mortality predictions at low dose, particularly between the LNT and quadratic theories, are significantly different. National Research Council, *The Effects on Populations to Low Levels of Ionizing Radiation,* BEIR III Report, National Academy Press, Washington, 1980.
12. At very high doses of the agent the risk of cancer induction decreases because the probability of killing cells exceeds the probability of inducing cancer.
13. *Supra* note 11.
14. Cancer incidence does not increase indefinitely with increasing dose. When the dose exceeds about 300 rad, cancer incidence begins to decline because tumor-forming cells are killed at high dose.
15. Pierce, D. et al., Studies of the mortality of atomic bomb survivors, Report 12, Part 1, Cancer, 1950–1990; *Radiation Research,* 146, 1, 1996.
16. Ibid. Supralinear dose response is also consistent with the Japanese atomic bomb survivor data.
17. Calabrese, E.J. and L.A. Baldwin, Toxicology rethinks its central belief: Hormesis demands a reappraisal of the way risks are assessed, *Nature,* 421, 691, February 13, 2003; Kaiser, J., Sipping from a poisoned chalice, *Science,* 302, 376, October 17, 2003.
18. All substances are poisons: there is none that is not a poison. The right dose differentiates a poison and a remedy (Paracelsus, 1493–1541).
19. National Research Council, *Health Risks from Exposure to Low Levels of Ionizing Radiation,* BEIR VII Report, National Academies Press, Washington, DC, 2005.
20. United Nations Scientific Committee on the Effects of Atomic Radiation, *Sources and Effects of Ionizing Radiation,* UNSCEAR, 1994 Report to the General Assembly, with Scientific Annexes, United Nations, New York, 1994; Mossman, K.L., Deconstructing radiation hormesis, *Health Physics,* 80, 263, 2001; Cardis, E. et al., Risk of cancer after low doses of ionising radiation: retrospective cohort study in 15 countries, *BMJ,* 331, 77, 2005.
21. Application of hormesis in risk assessment and risk management is also complicated by definitional problems and nonuniformity. Some investigators include treatment of cancer with low doses of radiation as hormesis when it is clearly a therapeutic application. In some studies hormesis was observed for some tumor types but not for others.
22. The data used in Figure 2.6 came from the following studies: Evans, R.D. et al., Radiogenic effects in man of long-term skeletal alpha-irradiation, *Radiobiology of Plutonium,* Stover, B.J. and Jee, W.S.S., Eds., The J.W. Press, Salt Lake City, UT, 431, 1972; Mays, C.W. and Lloyd, R.D., Bone sarcoma incidence vs. alpha particle dose, *Radiobiology of Plutonium,* Stover, B.J. and Jee, W.S.S., Eds., The J.W. Press, Salt Lake City, UT, 409, 1972; Rowland, R.E. et al., Dose-response relationships for female radium dial workers, *Radiation Research,* 76, 368, 1978.
23. Other factors contribute to risk uncertainty including the animal model used, dose range, experimental design and conditions, and biological end points monitored (type of tumor, observation period, etc.).

24. National Research Council. *Health Effects of Exposure to Low Levels of Ionizing Radiation,* BEIR V Report, National Academy Press, Washington, DC, 1990.
25. U.S. Nuclear Regulatory Commission, Radiological criteria for license termination, *Federal Register,* 62, 39058, 1997.
26. Mossman, K.L., The linear no threshold debate: Where do we go from here?, *Medical Physics*, 25, 279, 1998.
27. Fear of radiation-induced cancer or other health effects is one of several factors that might be considered by individuals who decline medical x-ray procedures and by pregnant women who elect to have abortions. For instance, women also decline to have mammography procedures because of the cost of the procedure or pain and discomfort. See Albanes, D. et al., A survey of physicians' breast cancer early detection practices, *Preventive Medicine,* 17, 643, 1988; Trichopoulos, D. et al., The victims of Chernobyl in Greece: Induced abortions after the accident, *BMJ*, 295, 1100, 1987.
28. Regulations have been relaxed when specific agents are delisted or approved for uses that avoid risks. Saccharin was first listed as a possible human carcinogen in 1977 based on no human data and very limited animal data. A classification review by the U.S. Department of Health and Human Services resulted in delisting of saccharin as a carcinogen in 2000. Thalidomide was banned in the 1960s for use by pregnant women to prevent morning sickness because it caused serious birth defects, but it is now an approved therapy for leprosy and is being investigated for use in treating AIDS, tuberculosis, and other illnesses including cancer. The Delaney Clause, passed by Congress in 1958, regulated pesticide residues as food additives. In 1996, Congress passed the Food Quality Protection Act repealing the Delaney Clause in part by redefining pesticides as nonfood additives.
29. *Chlorine Chemistry Council v. EPA* 206 F.3d 1286 (D.C. Cir. 2000). The court vacated the maximum contaminant level goal (MCLG) for chloroform of zero, set under the Safe Drinking Water Act based on LNT, because EPA itself had accepted new evidence of a dose threshold. The enforceable standard for chloroform in drinking water is the maximum contaminant level (MCL). The MCLG is not a regulatory limit but is used as a guideline for setting the MCL.
30. Lowrance, W.W., *Of Acceptable Risk: Science and the Determination of Safety,* William Kaufmann, Inc., Los Altos, CA, 1976. The idea of acceptable dose begs the question: Acceptable to whom? What may be acceptable to one individual or group may be unacceptable to another. Achieving consensus on an acceptable dose is difficult because impacted or interested stakeholders have different perceptions of costs, benefits, distributive justice, etc.
31. Risks appear to be better understood when expressed as frequencies rather than as percentages. See Gigerenzer, G., *Calculated Risks: How to Know When Numbers Deceive You,* Simon & Schuster, New York, 2002.
32. National Council on Radiation Protection and Measurements, *Uncertainties in Fatal Cancer Risk Estimates Used in Radiation Protection,* NCRP Report 126, NCRP, Bethesda, MD, 1997. The uncertainty analysis in this report applies to doses where excess cancers have been observed in Japanese survivors of the atomic bombs. At low doses uncertainties are increased significantly because of large dose extrapolations.
33. Tubiana, M., Dose-effect relationship and estimation of the carcinogenic effects of low doses of ionizing radiation: The joint report of the Academie des Sciences (Paris) and of the Academie Nationale de Medecine, *Int. J. Radiation Oncology Biol. Phys.*, 63, 317, 2005.

3 No Safe Dose

U.S. regulatory agencies use the linear no-threshold (LNT) theory as a basis for risk estimation and regulatory decision making. In approaching risk assessment in this way, two problems immediately emerge. First, there is no generally agreed-upon principles that can be used to select one predictive theory to the exclusion of other biologically plausible alternatives.[1] As pointed out in Chapter 2 and later in this chapter, theory selection is driven by the need for simplicity and for conservative risk predictions. However, the simplest and most conservative theory may not necessarily be the most desirable because of the limited range of predictions that can be made. Complex theories with several parameters offer a wider range of predictions.[2] Second, theoretical risk predictions at environmental or occupational exposure levels preclude meaningful decision making because risk uncertainties are so large. The problem of uncertainty in decision making is discussed in Chapter 4.

Chapter 2 discusses several theories that could be used in regulatory decision making, but LNT is preferred by standards-setting organizations because it is simple to use and has biological plausibility. In the 1970s, the U.S Environmental Protection Agency (EPA) considered a number of models and theories as a basis for risk quantification but selected LNT theory because there was a successful track record in its utility. The old Atomic Energy Commission, the predecessor organization to the U.S. Nuclear Regulatory Commission (U.S. NRC), had successfully used LNT theory in quantifying risks of cancer from exposure to radioactive strontium and radioactive iodine fallout from atmospheric nuclear weapons testing. Compared to most other plausible alternatives, LNT provides a conservative estimate of risk. If an agency is wrong using an LNT-derived risk estimate, it is likely to be wrong on the safe side by overestimating risk. Alternative theories (e.g., theories that predict a threshold) provide a less conservative estimate of risk. If a theory predicts a threshold that does not exist, then actual risks will be discounted.

This chapter focuses on LNT as the predictive theory of choice in risk assessment as used by the EPA, U.S. NRC, and other federal standards-setting agencies. The National Council on Radiation Protection and Measurements (NCRP) and the National Academies also use LNT as a cornerstone for their recommendations to standards-setting organizations.[3] Ionizing radiation is used as the model agent for discussions. Radiation studies in the 1920s first led to the idea that biological effects could be modeled by a linear dose-response function. Further, the application of LNT to risk assessment and standards setting was first established for ionizing radiation. Risk assessment for chemical carcinogens is essentially based on the radiation experience. Much of the fodder that has fed the LNT controversy during the past several decades has been based on radiation studies. Finally more human experience is available for ionizing radiation than any other single human carcinogen.[4] A brief historical overview is presented as a basis for discussing the underlying biological assumptions that are important in understanding the current LNT controversy.

LNT: THE THEORY OF CHOICE

The main attraction of LNT theory is that only a single parameter is needed to construct a dose-response curve and predict risk. All that is required is a statistically significant dose point and a ruler to draw a line through that point to the origin of the graph. A straight line is defined by two points on a plane where the origin (point at zero dose and zero health effects) is a measured point.[5] The slope of the line is a measure of the risk coefficient that can be applied at any dose. Figure 3.1 illustrates how the LNT dose response is constructed using data from a published study of female breast cancer incidence following radiation exposure. LNT theory predicts that the risk per unit dose (the risk coefficient) is the same at any dose. Accordingly, the risk coefficient calculated using data from direct observations made at high dose is the same risk coefficient used at small dose to make theoretical predictions of health effects. This application of LNT theory has been used to predict the risk of breast cancer in women undergoing screening mammography. At a typical breast dose of 1–5 mSv (0.1–0.5 rad), increased risk is not measurable, but a theoretical risk can be calculated using LNT theory.

LNT theory has an interesting history that is useful in understanding its underlying biological bases and its application in regulatory decision making. The first hint that the dose response for radiation-induced effects might be linear came from

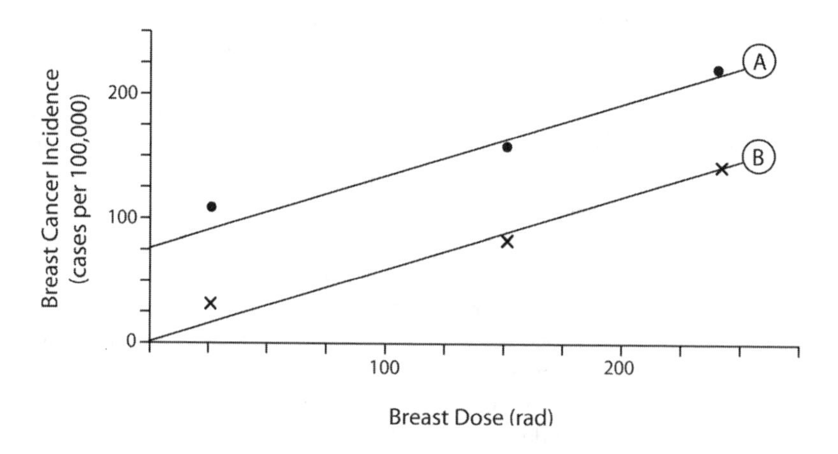

FIGURE 3.1 Linear no-threshold theory. Breast cancer incidence is usually represented by a linear dose response. Only two data points are required to determine the slope of the line. The slope (breast cancer incidence per unit radiation dose) is a measure of risk. Use of additional data points increases the accuracy of the slope estimate. Line "A" represents total breast cancer incidence (including contributions from radiation exposure); line "B" represents cancer incidence due to radiation exposure only. Breast dose is expressed in rad, the unit of absorbed dose of radiation. (Data from Boice, J.D. Jr., et al., Risk of breast cancer following low-dose exposure, *Radiology,* 131, 589, 1979.)

genetic studies carried out in 1927 by Hermann J. Muller.[6] Muller's work (awarded the Nobel Prize in 1946) clearly showed that ionizing radiation could cause heritable damage. Although his original data were not consistent with linearity, they did suggest the absence of a threshold dose.[7] Genetic mutations were considered until the 1950s to be the principal effects at small doses of radiation.[8] Nikolai Timofeeff-Ressovsky and other experimental geneticists advanced Muller's work during the 1930s.[9] Timofeeff's principal contribution was his observation of a linear relation between the total radiation dose and the number of mutations produced. He found that the yield of mutations did not change irrespective of whether the dose was administered in a single acute shot, given continuously at low levels for a long period of time, or given in several discontinuous fractions. His studies also confirmed Muller's observation of no-threshold dose. Timofeeff's analysis of the mutagenic properties of x-ray gave birth to what is now referred to as the target model of radiation action (Figure 3.2). Timofeeff thought that radiation damage was analogous to bombs hitting targets.[10] At the time, the idea of a "target" was strictly a biophysical construct. Now it is well known that the principal target is DNA.

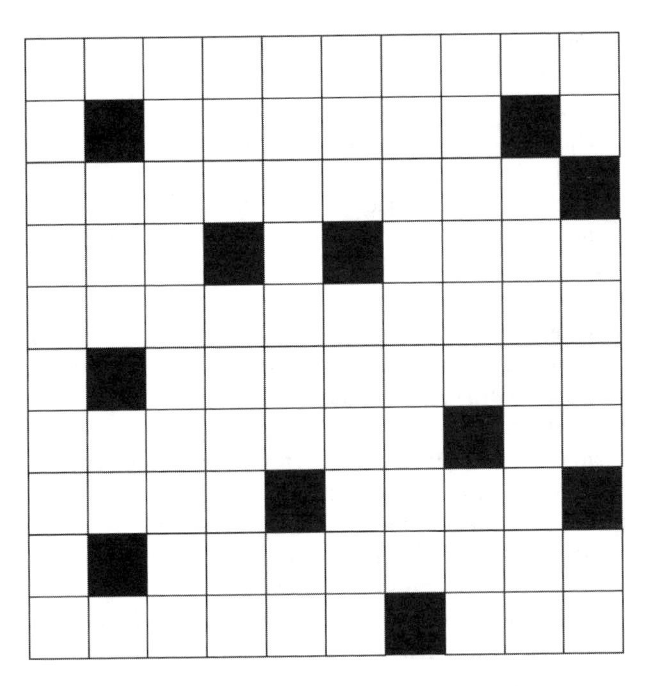

FIGURE 3.2 Target model of radiation action. The underlying biophysical basis for LNT theory is the target model. Humans and other complex organisms may be thought of as a collection of cells depicted as squares in the figure. When exposed to radiation cells are damaged (hit) in a random fashion as shown by darkened squares. For large cell populations and small doses, LNT predicts that the number of cells hit varies linearly with dose; doubling the dose doubles the number of cells hit. The target model assumes that cells are incapable of repairing damage and that cells behave autonomously and do not communicate with each other. These assumptions are now known to be incorrect.

The early studies of the genetic effects of radiation led directly to the development of LNT theory as the foundation of current radiation standards and radiation protection practice, the "no safe dose" concept, which has been the cornerstone philosophy of many antinuclear groups, and the target model, which for many years was the principal model for describing cellular effects of radiation. Timofeeff's observations of lack of a dose rate effect have also had important implications for radiation protection. The influence of protracted or fractionated exposure regimens on the biological effectiveness of a given total dose of radiation, particularly regarding cancer induction, is still unclear, but it is recognized as an important variable in risk assessment at low doses.[11] Timofeeff interpreted the lack of a fractionation or protraction effect as indicative of the lack of repair of radiation injury of "targets." Once a target is "hit" by radiation, it is permanently damaged. However, substantial scientific evidence has accumulated since the target model was formulated that repair of radiation injury does occur.

Interest in health effects of radiation shifted from concerns about genetic effects to cancer in the 1950s. A link between radiation and cancer was not entirely surprising. There was substantial evidence from radium dial worker studies in the 1920s and from radiation therapy experiences that cancer could occur but at relatively high doses (e.g., > 500 mSv).[12] Furthermore, the early radiation genetics studies suggested that cancer (believed to be a result of mutations) could be caused by radiation because of its mutagenic properties. In the 1950s epidemiological studies of the Japanese survivors of the atomic bombings were ongoing, and reports in the medical literature documented small excess cancer risks in patients undergoing high-dose radiation therapy for benign disease. These early reports subsequently led to the study and evaluation of a large number of chemical agents for their mutagenic and carcinogenic properties.[13] E.B. Lewis first recognized that numbers of cancers could be predicted using LNT theory.[14] For example, if 1,000 cancers occur as the result of exposure of a population of 1,000,000 persons to an average dose of 0.1 Sv (sievert), then 100 cancers would be expected in this same population had the average exposure been 0.01 Sv. Lewis also introduced the concept of collective dose. This is the sum of all individual doses in the population at risk (it is numerically equal to the product of the average population dose and the size of the population) and is measured in terms of person-dose such as person-Sv. Linearity predicts that the number of radiation-induced cancers is the same for a population of 1,000,000 exposed to an average dose of 0.1 Sv and a population of 100,000 exposed to an average dose of 1 Sv. In each case the collective dose is 100,000 person-Sv. Recently the International Commission on Radiological Protection (ICRP) and other authoritative bodies have considered recommendations to limit the utility of collective dose in radiation protection because of serious misuse of the concept. Examples of misuse are described later in this chapter.[15]

LNT theory is hard to argue against. Its simplicity and inherent conservatism are attractive to decision makers and policy makers. From a precautionary approach, LNT is the most appropriate strategy for estimating risk in populations exposed to very small doses where there is no direct evidence of risk.[16] But LNT may overestimate risk if the underlying dose response in non-linear. Compared to other more complex theories (see the discussion of other theories in Chapter 2), LNT is easier to communicate to the lay public. From a biophysical perspective, LNT theory reflects single order kinetics—that is, only a single event, either one photon or one molecule, is needed to produce the effect. This "straw that broke the camel's back" perspective argues that since the body constantly accumulates damage by absorption of photons and chemicals from the environment, the next photon or next molecule is all that is needed to produce the effect. In this context, LNT makes a good deal of sense if the causal agent is both necessary and sufficient (i.e., it is the only requirement for the disease). The idea is that during life incremental damage is accumulated from various sources, including normal cellular metabolic activity, natural background radiation, and environmental pollutants. At some point all that is needed is one more insult, and that may come from a medical x-ray or from exposure to some other cancer-causing agent. What is problematic with this line of reasoning is that the body is bombarded by natural sources of carcinogens all of the time. The single photon or molecule that is responsible for disease is more likely to come from carcinogens encountered in the course of everyday life.

Molecular biology and molecular genetics studies suggest that carcinogenesis is a very complex pathologic process with multifactorial features involving the interplay of genetic and environmental factors. Our current understanding of cancer development would suggest that a single photon or a single molecule is not sufficient, in itself, to cause cancer. To say that a single photon can result in cancer ignores the importance of other host and environmental factors that contribute to risk. As shown in Figure 3.3, carcinogenesis involves a sequence of genetic changes in cells. Colon cancer is an example where a number of sequential mutations and cellular changes occur over a long period of time. What is important is that mutational damage is accumulated over time, not the specific sequence of mutational events that occurs.

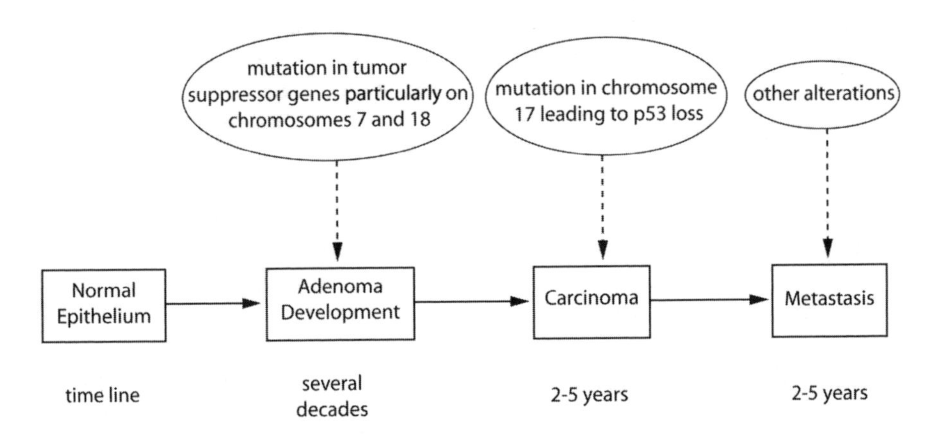

FIGURE 3.3 Cancer is a multistage process. This schematic illustrates the multistage process characteristic of colon cancer, a very common form of cancer in humans. Cancer does not arise as a result of a single genetic event in cells. Cancer develops in stages and arises from alterations in multiple genes (as illustrated by the mutations that arise at different stages of colon cancer development). These changes lead to pathologic changes in the colon epithelium (the functional tissue within the colon), from a normal epithelium, to early and late adenoma, to early and late cancer. These changes usually take many years to develop. (Modified from Fearon, E.R. and Vogelstein, B., A genetic model for colorectal tumorigenesis. *Cell,* 61, 759, 1990.)

The multifactorial nature of carcinogenesis suggests that LNT may be too simplistic. Cancer is more than damage to individual cells. Cancer is a disease of tissues that involves complex cell and tissue interactions. Such interactions cannot be easily modeled by LNT theory.

LNT theory predicts that damage is additive. Thus, net damage resulting from a dose administered over a period of time would be predicted to be the same as the net damage resulting from the same dose delivered instantaneously. It is now known that additivity does not strictly hold. Cells, tissues, and organs are able to "repair" damage such that chronic exposures are less biologically effective than acute exposures. Regulatory and policy decision makers now apply an external correction factor to LNT-derived risks to account for repair when radiation doses are administered over an extended period. The dose and dose-rate effectiveness factor (DDREF) simply changes the slope (i.e., risk coefficient) without changing the shape of the dose-response curve. Values for DDREF are not known very well; accordingly, use of the correction factor introduces considerable uncertainty in risk assessment.[17]

LNT theory also predicts that cells respond autonomously to a dose of radiation. The target model, the biophysical basis of LNT, requires that the probability of damaging one cell is independent of the probability of damaging a neighboring cell. At low dose LNT theory predicts that the number of cells damaged is doubled if the dose of radiation is doubled.[18] Recent laboratory studies suggest that the principle of independence is not valid. Studies exploring extracellular signaling suggest that cells are not necessarily the masters of their own fate. One extracellular phenomenon, termed the bystander effect, posits that irradiation of one cell can influence the response of surrounding, unexposed cells. How cells communicate is not well understood at this time, but it is clear from these studies that radiation responses in the multicellular organism are very complex and cannot be modeled simply by looking at cellular responses. Understanding how radiation damage may be modified by the tissue microenvironment is important in assessing risk.[19]

Extensive studies of cancer induction in humans and experimental animals indicate that the shape of the dose response is cancer site dependent. Leukemia appears to follow a curvilinear dose response; bone cancer appears to be associated with a dose threshold; breast cancer and thyroid cancer are consistent with LNT theory. Clearly, diversity of dose response is problematic for risk assessment. Although LNT provides a conservative estimate of risk, application of LNT theory may overestimate risks for cancer types characterized by nonlinear dose responses.

Whether bystander effects, repair, and other biological processes are significant modifiers of low-dose risks is unclear. Modifying a risk of the order of 1 in 10,000 (the lifetime cancer mortality risk associated with some types of diagnostic x-ray procedures) is not easily measurable by standard epidemiological methods. Some risk modifiers may serve to cancel each other out or to amplify effects, further complicating the problem. Repair of radiation damage will tend to reduce risk, but bystander effects and genomic instability may have the opposite effect by increasing risk. The problem is almost intractable because risks at environmental and occupational dose levels are very small to begin with, and any variations in risk are difficult to measure.

THE LNT CONTROVERSY

The LNT debate has been ongoing for more than three decades. Although LNT is a simple, straightforward theory, its implementation has caused serious practical problems in radiological health and safety. The notion that any dose, no matter how small, may cause cancer has led to the widespread belief that no dose is safe. As a consequence huge sums of money have been allocated to reducing radiation doses to levels as close to zero as possible. In 1966, ICRP noted that the assumptions of no additivity and no threshold may be incorrect but could not see a practical alternative to LNT theory that is unlikely to underestimate risk.[20] Most authoritative bodies continue to support LNT theory in regulatory decision making and as a foundation for radiation protection practice, but some prominent organizations such as the Health Physics Society and the French Academy of Sciences have taken a more critical position.[21]

LNT theory has come under attack on purely scientific grounds. Analysis of a number of epidemiological studies of the Japanese survivors of the atomic bombings and workers exposed to low-level radiation suggest that the LNT philosophy is overly conservative, and low-level radiation may be less dangerous than commonly believed. Proponents of current standards argue that risk conservatism is justified because low-level risks remain uncertain, and it is prudent public health policy; LNT opponents maintain that regulatory compliance costs are excessive, and there is now substantial scientific information arguing against LNT theory.[22]

Many questions about the existence of a threshold and the shape of the dose-response curve at low doses and dose rates remain unanswered. By retaining LNT theory as a philosophical foundation, radiation protection has been operating for more than 30 years on the basis of cautious empirically unproven assumptions. The obvious question in the LNT debate is: "If not LNT, then what?" Alternative predictive theories, such as the linear quadratic or pure quadratic theories also have significant uncertainties because of the paucity of statistically significant data in the low-dose range. Selecting a particular predictive theory to the exclusion of biologically plausible alternatives may be an intractable problem.

In a risk-based decision framework, policy makers and regulators are in a serious bind to select the most appropriate theory for risk assessment and risk management. Unfortunately, current scientific evidence does not provide a clear path. The limitations of LNT theory are fully appreciated by scientists and decision makers, but there may be no better alternative. Authoritative bodies such as ICRP and NCRP continue to support LNT because it is a reasonable compromise position among alternative theories. LNT is also intelligible to the nonscientist, and this helps to keep risk assessment transparent to the public.

In his 1996 Sievert Lecture, Dan Benninson expressed the view shared by many leaders in radiological protection that LNT theory has almost reached the status of scientific dogma. Benninson's lecture was an emotional plea to reject competing theories.[23] Such pleas do a disservice to radiation protection and public health by stifling scientific discourse. Radiation-protection authorities force scientific justification of LNT when such justification is difficult to defend. The decision to use LNT theory is a political one and should not be argued on scientific grounds. Scientific evidence is not robust enough to exclude any predictive theory at small doses.

Elements of the Debate

Three major issues have driven the LNT debate: First, regulatory compliance costs too much. Regulations depend on the concept that reduction in dose leads to a concomitant reduction in health risk. However, there is little epidemiological evidence (except for the reduction in lung cancer in populations following reduction in cigarette consumption) to show that reducing exposure to environmental carcinogens leads to a reduction in cancer risk. Second, support of LNT theory and the idea that any radiation dose is potentially harmful has resulted in a public relations nightmare for nuclear technologies (including medical applications of radiation). The nuclear community's unfailing support of LNT theory has made it almost impossible to respond effectively to alarmists' claims that any dose of radiation is dangerous. The resulting public outrage from dangerous radiation has led to overregulation of nuclear industries, resulting in billions of dollars in compliance costs. Third, based on a growing body of scientific evidence, LNT theory appears to be an oversimplification of the dose-response relationship and may overestimate health risks in the low-dose range.

The Question of Thresholds

Threshold supporters argue that a threshold dose for carcinogenesis must exist because DNA damage (the key initiating event in cancer induction) occurs at a high rate naturally, even in the absence of radiation exposure, and that a small dose of radiation (e.g., 10 mSv) increases the total number of DNA base lesions and single strand breaks by an insignificant amount. Approximately 150,000 single-strand breaks and nitrogenous base lesions occur spontaneously in every mammalian cell per day; a daily dose of 10 mSv adds only 20 additional events per day.[24]

LNT proponents argue that base lesions and single-stranded DNA lesions occur frequently, and are readily repaired by the cell. However, small doses of ionizing radiation significantly increase the burden of double-stranded DNA lesions.[25] The spontaneous incidence of DNA double-strand breaks is approximately 0.01 per cell; 10 mSv results in a 40-fold increase in the frequency of these lesions. These events are less efficiently repaired and have more serious consequences for the cell, including oncogenic transformation. At small doses (e.g., 10 mSv), DNA-dependent cell damage is a linear function of dose. However, this does not mean that cancer incidence/mortality is also a linear function of dose since carcinogenesis is a multifactorial pathological process.[26] Transformation of cells is the first step in carcinogenesis; damaged cells must undergo further changes related to promotion and progression as a prerequisite to the clinical appearance of disease. Ionizing radiation can influence post-initiation events in carcinogenesis.

Repair of Radiation Damage and Cellular Autonomy

LNT theory predicts that radiation damage is cumulative and that there is no biological repair of radiation damage. Accordingly, health effects should be independent of dose rate. If radiation injury is repairable, then low dose rate exposures should be less biologically effective than high dose rate exposures.[27] Although "repair" (in the context of reduced risk at low dose rate) has been convincingly demonstrated in cell culture systems and to a lesser degree in human populations through epidemiological studies, the importance of repair as a determinant of radiogenic cancer risk is still unclear. Much work remains to fully understand repair of radiation damage. Repair has been clearly documented at high doses (usually greater than 1 Sv) where biological responses can be readily measured. At dose levels of relevance in radiation protection (usually <10 mSv), radiogenic effects are difficult to observe. Accordingly, repair at these dose levels is not likely to be an important determinant of risk.

The biophysical argument for linearity also includes the assumption that cells behave autonomously, i.e., damaged cells do not affect other cells. Cancer is a complex disease that involves the interaction of cells. Proponents of linearity argue that such multicellular interactions should not affect linearity as long as the rate-limiting step is a single-cell process.[28] However, there are now cooperative effects including the bystander effect and genomic instability that call into question the autonomy assumption. Nontargeted effects are discussed in greater detail in Chapter 7.

USES AND MISUSES OF LNT

LNT and collective dose have utility in several areas of risk assessment and risk management. Comparing collective doses is a useful quantitative measure of the magnitude of exposure to a particular radiation source or group of sources in a defined population. A reference collective dose must be chosen carefully in order to make the comparison valid. The collective dose from natural background radiation may be a useful reference when evaluating exposures to a large population exposed to small doses. Comparing collective doses from different exposure scenarios allows for a quantitative measure of severity of exposure. Temporal trend analysis of collective dose and setting specific and realistic collective dose "goals" may be useful in evaluating the effectiveness of risk-management programs. It should be emphasized that a zero goal is not realistic since it cannot be expected that doses to all workers or to individuals in a population at risk can be reduced to zero. Using LNT and collective dose in these ways avoids the pitfalls and problems associated with risk calculation, particularly when very small doses are involved as typically encountered in environmental and occupational exposure settings.

Difficulties surface when attempting to relate exposure to small doses with health risk using LNT theory. LNT predicts that any dose is associated with risk. But in reality apparent thresholds exist such that risks are unobservable at small doses. Is there a difference, from a public health perspective, between zero risk and a risk that cannot be measured? The following examples illustrate problems encountered when LNT and collective dose are misused in efforts to calculate health risks and public health impacts. The result of this misuse is that health risks are exaggerated, result and the idea that very small doses of radiation may be harmful endures. The first case is an illustration of the misapplication of LNT to predict health effects involving very large populations and very small radiation doses. The second case is an example of the misuse of the LNT theory to predict population risks based on calculated individual risks. The third case illustrates the problems in interpreting risks based on extrapolating to very small doses using the LNT theory.

Case 1: Estimation of Health Effects of Fallout from the Chernobyl Reactor Accident

In April 1986 the nuclear power plant at Chernobyl sustained serious damage accompanied by a massive atmospheric release of radioactive material. The accident resulted in the loss of 31 rescue and firefighting personnel, but thousands of others were exposed to subacute doses of radiation that increased the probability of cancer. The radioactive fallout from the Chernobyl accident was measured all over the Northern hemisphere, including the United States and Canada. Estimates have been made to determine the health impact of the radioactivity released from the Chernobyl reactor accident. In one study average doses to the U.S. and Canadian populations were estimated.[29] Using cancer mortality risk estimates derived from the Japanese atomic bomb survivor data, the number of expected cancer deaths in the U.S. and Canada were calculated. This risk-assessment exercise is an example of applying the collective dose concept incorrectly. In this case a very small average dose to a large population is used to obtain a small collective dose to the population. The U.S. and Canadian estimate turns out to be 20 excess cancer deaths. This number is impossible to verify because the actual doses received were exceedingly small. More importantly, it is not possible to detect such a small number of cancers in face of the estimated 48 million cancer deaths expected from all causes of cancer. Radiogenic cancers cannot be distinguished clinically from spontaneous cancers or cancers resulting from other known causes. The average dose to Americans and Canadians in this assessment is theoretically derived and cannot be measured directly. The individual dose estimates are so low that they are within the statistical variations in natural background radiation levels in North America. An estimated number of cancer deaths is obtained in the situation only because of the very large population involved. In applying collective dose and using LNT to estimate risk, it is important to recognize the limitations in the calculations. Doses in the U.S. and Canada as a result of the Chernobyl accident have no real meaning because they are so small. Furthermore, no epidemiologic studies or evidence are available to suggest that such small doses may be harmful. A more appropriate approach to risk assessment in this case would be to compare population doses with natural background radiation levels. In a risk context, no detrimental health effects have been observed at radiation levels approaching several times natural background radiation levels. According to the U.S. National Research Council's BEIR V Committee, at such doses and dose rates, the lower limit of the range of uncertainty in radiogenic risk estimates extends to zero.[30]

Case 2: Childhood Cancer Following Diagnostic X-Ray

Since the introduction of computerized tomography (CT) in the 1970s, CT examinations to diagnose a wide spectrum of diseases have rapidly increased in frequency particularly among pediatric patients. In the U.S. approximately 600,000 abdominal and head CT examinations are performed annually in children under the age of 15 years.[31] CT is an extremely valuable tool for diagnosing illness and injury in children. For an individual child, the risks of CT are small, and the individual risk-benefit balance almost always favors the benefit. Since children may receive higher doses of radiation than adults (because of their smaller size) and because children are generally more sensitive to the effects of radiation than adults, there is concern that the large number of pediatric CT examinations may result in a significant increase in radiation-induced cancer. Although individual risks may be small, there is a large population exposed and that, according to one Columbia University study, is what makes pediatric CT a public health problem.[32] This public health issue has received widespread attention in both the popular and medical press. The study suggests that pediatric CT will result in a significant increase in lifetime radiation risk because of the high doses in pediatric patients. The report estimates that 500 cancer deaths will occur as a result of the 600,000 pediatric CT examinations performed annually. The basis for this conclusion is an application of LNT theory. Although doses from pediatric CT examinations can be quite high (particularly if multiple examinations are conducted), the doses received to critical organs such as the bone marrow and thyroid gland are not so high that a serious radiological risk is present. Risks to the individual pediatric patient are generally considered so small that they are almost always outweighed by the benefits of the clinical procedure.

However, the study authors argue that because a large population of patients is involved, there is a *public* health problem. Using LNT, a theoretical estimate of the number of cancers likely in the population at risk may be calculated. There are two problems with this approach. First, the public health metric is the product of the individual patient risk and the size of the population. However, the individual risk is independent of the population size. A small probability when multiplied by a sufficiently large population yields a finite population risk. It is on this basis that small doses are considered potentially harmful. However, the fallacy in this reasoning is that population risks are mapped on individual risks. If everyone in the population receives the same dose, individual risks are the same for every member (assuming homogeneity of responses) regardless of the size of the population. The risk to the individual is not relevant to the population risk. Second, even if the calculation of number of cancer deaths was proper, expression of the public health risk as a single number (500 deaths) is inappropriate and reflects a level of scientific certainty that is not supported by the underlying science. At the very least such risk assessments should be presented as a range of possible outcomes. It is possible that no cancers may be produced as a result of such exposures since there have been no epidemiological studies conducted to date that suggest that cancers in children are likely to occur following doses comparable to that received in pediatric CT examinations.[33]

The Columbia University study has served a useful purpose by heightening the Radiology community's awareness to reduce radiation doses in pediatric CT studies. There is concern that radiation doses from CT scans are unnecessarily high, especially since the frequency of CT examinations is increasing in many countries.[34] Because individual doses are small, radiation risks cannot be quantified, and numbers of cancer deaths expected cannot be estimated in a meaningful way. Nevertheless it is important to reduce dose when it is possible to do so without compromising diagnostic quality. Arguing that hundreds of cancer deaths may result from pediatric CT studies unnecessarily raises public fears and concerns and unfortunately perpetuates the notion that any dose of radiation is harmful. Such fears may result in some children being denied necessary CT studies and compromising health care because of the fear of a theoretical cancer risk. This does not mean we should use CT studies injudiciously. On the contrary, such studies should be conducted only when there is a clinical justification. Care should be taken to optimize radiation dose and reduce the frequency of repeat examinations.

Case 3: Public Health Impacts from Radiation in a Modern Pit Facility

The U.S. Department of Energy (DOE) National Nuclear Security Administration (NNSA) proposes to construct a Modern Pit Facility (MPF) to address a critical gap in the long-term nuclear readiness of the United States. NNSA is responsible for the safety and reliability of the U.S. nuclear weapons stockpile, including production readiness required to maintain that stockpile. Since 1989 (the year the Rocky Flats Plant in Colorado closed), the DOE has been without the capability to produce stockpile-certified plutonium pits.[35] The proposed MPF will process old pits, manufacture plutonium components, and assemble complete pits of current or new design for use at the Pantex Plant near Amarillo, Texas.

As part of the planning process, NNSA drafted an Environmental Impact Statement (EIS) to address, among other things, the environmental and public health impact of an MPF.[36] According to EIS calculations, radiological impacts to workers and the public would occur during normal MPF operations. LNT theory was used to predict the number of cancer deaths expected among workers and the general public. For the estimated 1,000 MPF workers in a 450-pits-per-year facility, one radiogenic fatal cancer would be expected to occur every 4,900 years of operation; for members of the general public living near the facility (the size of the population depends on where the MPF is sited), one fatal cancer has been estimated for every 1.5 billion years of operation. The use of LNT to predict health impacts in this situation is problematic and illustrates practical limitations of LNT as a predictive theory. First, although the worker population may be well defined, it is too small to provide any meaningful estimation of radiogenic effects because the radiation risk is so small and the spontaneous cancer mortality rate (about 20%) is so high. In the case of public impact, the average dose to any member of the public living near the plant would be expected to be no more than 1% of the average annual natural background radiation level that everyone is exposed to. Using LNT to extrapolate to such small doses has no real meaning particularly when expressed as cancer fatalities per number of years of plant operation. One cancer among workers per 4,900 years of operation or one cancer in the general population per 1.5 billion years of operation essentially means that no radiological health impacts are expected. Because of the large uncertainties in risk at very small doses, it is prudent to express risk as a range of possible outcomes, including the possibility of zero health effects.

LNT CONSEQUENCES

Use of LNT has had serious consequences with regard to public perception of small risks and economic and social costs of managing very small risks. Over time, the idea that a small amount of radiation may cause harm has led to the belief that radiation at any dose will cause harm and that no dose of radiation can be considered safe. Special interest groups (e.g., antinuclear activists) are not the only ones guilty of this. The media, government regulators, and the nuclear industry (including medicine) are also to blame. By basing regulatory decision making on LNT theory, government regulators unwittingly support the philosophy that no dose is safe. The idea that no dose is safe is perfect fodder for print and broadcast journalists to distort and sensationalize stories about the hazards of low-level radiation in the interest of selling papers. The Chernobyl accident in 1986 resulted in serious public health and environmental effects in Western Russia and nearby countries. However, reports of public health effects in North America as a result of the Chernobyl fallout can only be labeled as sensational and ludicrous. Fallout levels in the U.S. were barely detectable. Dose calculations to the U.S. population turned out to be a small fraction of annual natural background radiation levels and would not result in any increase in cancer.

Fear of small doses of radiation has had a serious long-term impact on how technological risks are perceived by individuals and society. Risk-avoidance behavior has led to fear-driven technological decisions rather than science-based ones. LNT theory continues to be used to estimate health effects at dose levels associated with risks that are too small to measure reliably. Use of LNT in risk assessment has evolved to the point where we fail to distinguish between theory and reality. Predictive theories (including LNT) do not provide us with actual dead body counts. But some (including antinuclear people, regulators, and the media) ignore the limitations of predictive theories. The public has been brainwashed into believing that any dose of radiation is harmful, and has been made to fear even tiny doses, far below natural background radiation levels. Such fears have had a widespread impact on how individual and societal decisions are made regarding nuclear technologies. We will look at three examples.

No new nuclear power plants have been approved for construction in the U.S. since the Three Mile Island nuclear accident near Harrisburg, PA, in 1979. This situation persists in spite of the fact that nuclear power generation is generally regarded as safe, and future design modifications will make operating plants even safer. It is an important carbon-free source of power that can potentially make a significant contribution to future electricity supply while helping to address the global warming problem. Although fear of low-level radiation risks continues to be a major concern regarding the future of nuclear power, questions also remain about the adequacy of plant safety systems, disposal of spent fuel, and security of nuclear plants.

Radioactive waste disposal continues to be a political problem with no end in sight. In 1980, the U.S. Congress passed the Low-Level Radioactive Waste Policy Act (LLRWPA), which was signed into law by President Jimmy Carter. The act was a blueprint for a national plan to establish regional radioactive waste sites in order to create regional equity for the disposal of low-level waste. By almost any measure, the legislation has failed. Radioactive waste at nuclear power plants must be stored on site. Smaller waste generators (e.g., hospitals and medical centers) store large quantities of waste on site because of limited off-site disposal options. More than 20 years after passage, the LLRWPA has produced 10 interstate compact commissions and no new disposal facilities. The Southwest Compact and other compacts created by the LLRWPA have yet to build regional disposal facilities. Concerns are primarily political rather than scientific and technical. Even when suitable sites (such as Ward Valley in California) are identified and subjected to extensive scientific review and risk analysis, special-interest groups have successfully argued that such sites are not safe because there is the very unlikely probability of loss of containment with seepage of waste into the environment. These special interest groups have convinced public officials that resulting small public doses are harmful even though risks can only be determined theoretically. Assuring adequate access to waste disposal facilities will require the political will of the federal government. Congress must recognize that the LLRWPA has failed and that the nation does not need 10 new disposal facilities. This is a national problem requiring a national solution.[37]

Medical applications represent the largest anthropogenic source of radiation. Over 75% of the U.S. population gets medical or dental x-rays annually.[38] Clinically justified x-ray procedures are important in disease diagnosis and therapy. Yet some people refuse diagnostic x-ray procedures because of fear of radiation. By refusing needed tests, patients may be compromising medical management of potentially serious conditions. The issue is, however, more complex than simply fear of small risks. Patients also refuse tests because they are uncomfortable or painful (e.g., mammography), and in some cases the tests are not covered by insurance, thus requiring substantial out-of-pocket costs. Asymptomatic patients may be unwilling to undergo tests because they do not want to learn that they have cancer or some other serious disease when they feel perfectly alright.

Distorted views of low-level health risks lead to irrational decision making and the expenditure of large sums of money to reduce risks that are already very small to begin with. Although scientific and technological perspectives are important in evaluating technological risks, they should not be the sole drivers in public decision making. It is imperative that the public evaluate technological risks realistically based on sound science, but nontechnical issues (e.g., equitable distribution of risks and benefits of technology) and perspectives of interested and impacted stakeholders must also be taken into consideration by decision makers. Clearly risk perception is a major obstacle to technological progress. The stagnation of nuclear power since the 1970s is a perfect example. Believing that low-level radiation is harmful and other misperceptions can lead to diversion of limited resources away from managing really significant risks to the public health and the environment. Efforts to address determinants of risk perception are important to assure that societal decisions are made wisely and for the benefit of all.

NOTES AND REFERENCES

1. This is in contrast to the situation with product liability where the Supreme Court decision in *Daubert v. Merrill Dow* 509 U.S. 579, 1993 clearly establishes rules for scientific evidence.
2. Theories with many parameters are problematic because almost any outcome is possible. The ideal theory should have the minimum number of parameters with biological meaning. The late mathematician John von Neumann is claimed to have remarked that with four parameters he could fit an elephant and with five he could make him wiggle his trunk.

3. National Council on Radiation Protection and Measurements. *Evaluation of the Linear Nonthreshold Dose-Response Model for Ionizing Radiation.* NCRP Report No. 136, National Council on Radiation Protection and Measurements, Bethesda, MD, 2001; National Research Council, *Health Risks from Exposure to Low Levels of Ionizing Radiation,* BEIR VII Report, National Academies Press, Washington, DC, 2005.
4. Substantial human epidemiological data are also available for cigarettes, but cigarette smoke is a collection of thousands of chemicals, many of which are poorly understood as carcinogens.
5. This point is determined by measuring the spontaneous incidence of cancer deaths or other health effects (in the absence of radiation exposure). All other data points at doses greater than zero are corrected for the spontaneous incidence rate.
6. Muller, H.J., Artificial transmutation of the gene, *Science,* 66, 84, 1927.
7. A threshold may be defined as the dose below which no effects or responses are observed.
8. Small doses may be defined as less than 100 mSv (consistent with a risk of about 1 in 1,000, defined as low risk in Chapter 1). Examples include exposure to cosmic radiation from round-trip airline trip between New York and London (0.1 mSv); annual U.S. natural background radiation excluding contributions from radon (1 mSv); annual U.S. natural background radiation including radon (3 mSv); mammography screening examination (3 mSv); annual radiation worker exposure limit (50 mSv).
9. Readers interested in a detailed account of the history of radiation genetics, particularly from 1927 through the 1950s, are referred to the following technical reviews: Muller, H.J., The nature of the genetic effects produced by radiation. In Hollaender, A., Ed., *Radiation Biology,* McGraw-Hill Book Company, Inc., New York, 351, 1954; Muller, H.J., The manner of production of mutations by radiation. In Hollaender, A., Ed., *Radiation Biology,* McGraw-Hill Book Company, Inc., New York, 475, 1954; Russell, W.L., Genetic effects of radiation in mammals. In Hollaender, A., Ed., *Radiation Biology,* McGraw-Hill Book Company, Inc., New York, 825, 1954; Sobels, F.H., Radiation genetics, foundation and perspectives. In Duplan, J.F. and Chapiro, A., Eds., *Advances in Radiation Research,* Volume 1, Gordon and Breach Science Publishers, New York, 277, 1973; Carlson, E. A., *Genes Radiation and Society: The Life and Work of H. J. Muller.* Cornell University Press, Ithaca, NY, 1981.
10. Timofeeff-Ressovsky, N.V., Zimmer, K.G., and Delbruck, M., Uber die natur der genmutation und der genstruktur. Vierter teil: Theorie der genmutation und der genstruktur, *Nachr Ges Wiss Gottingen Math Physik Kl Fachgruppe VI,* 1, 156, 1935.
11. National Research Council, *Health Risks from Exposure to Low Levels of Ionizing Radiation,* BEIR VII Report, National Academies Press, Washington, DC, 2005.
12. Mossman, K.L. A brief history of radiation bioeffects. In *Health Effects of Exposure to Low-Level Ionizing Radiation,* Hendee, W.R., and Edwards, F.M., Eds., Institute of Physics Publishing, Philadelphia, PA, 1996.
13. A number of agents has now been identified as cancer-causing or suspected of being cancer-causing agents in humans. See U.S. Department of Health and Human Services, Public Health Service, National Toxicology Program, *Report on Carcinogens,* 11th ed., 2005; International Agency for Research on Cancer (IARC), *Overall Evaluations of Carcinogenicity to Humans,* http://monographs.iarc.fr/monoeval/crthall.html (accessed March 2006).
14. Lewis, E. B., Leukemia and ionizing radiation, *Science,* 125, 965, 1957.

15. For a detailed review of collective dose and its limitations see National Council on Radiation Protection and Measurements, *Principles and Applications of Collective Dose in Radiation Protection,* NCRP Report No. 121, Bethesda, MD, NCRP, 1995.
16. Supralinear theories that predict higher risk at small doses are more conservative than LNT. However, supralinear theories are not typically used in regulatory decision making.
17. *Supra* note 11. The National Academies' BEIR VII Committee took a very conservative position and assumed a DDREF of 1.5 in its solid cancer risk estimates recognizing the substantial uncertainty in the correction factor. See also National Council on Radiation Protection and Measurements, *Uncertainties in Fatal Cancer Risk Estimates Used in Radiation Protection*, NCRP Report No. 126, Bethesda, MD, NCRP, 1997.
18. At low dose the probability that a cell will receive two or more hits is negligible.
19. Multicellular responses through extracellular signaling are important risk determinants. Bystander effects (unexposed cells are affected by irradiated neighbor cells) and genomic instability (cells surviving radiation exposure produce progeny that have unstable DNA that is unlikely to be the result of conventional mutation) appear to be two key extracellular communications phenomena. For an overview of extracellular signaling and cancer risk see Barcellos-Hoff, M. and Brooks, A.L., Extracellular signaling through the microenvironment: A hypothesis relating carcinogenesis, bystander effects, and genomic instability, *Radiation Research,* 156, 618, 2001.
20. International Commission on Radiological Protection (ICRP). *Recommendations of the International Commission on Radiological Protection.* ICRP Publication 9, Pergamon Press, Oxford, U.K., 1966.
21. The ICRP supported LNT theory as a basis for its 1990 recommendations (see International Commission on Radiological Protection [ICRP]. *1990 Recommendations of the International Commission on Radiological Protection.* ICRP Publication 60, Pergamon Press, Oxford, 1990); The ICRP re-affirmed its position in support of LNT with the publication of ICRP Publication 99 in 2005 [see International Commission on Radiological Protection, *Low-Dose Extrapolation of Radiation-Related Cancer Risk*, ICRP Publication 99, Annals of the ICRP 35(4), 2005]. The National Radiological Protection Board (NRPB) also reaffirmed the use of LNT theory in radiation protection in its 1995 review of radiogenic risks at low doses and low-dose rates (National Radiological Protection Board [NRPB], *Risk of Radiation-Induced Cancer at Low Doses and Low Dose Rates for Radiation Protection Purposes*, Documents of the NRPB, Volume 6 [1], NRPB, Didcot, 1995). Central to the NRPB position was the conclusion that, although repair of radiation injury has been well documented in laboratory investigations, there is little human epidemiological evidence to suggest that repair at low dose influences radiogenic risk. The Health Physics Society has questioned the validity of the LNT theory in the range of occupational and environmental exposures of about 1–10 mSv because of the lack of health effects data below 100 mSv (Mossman, K.L. et al., Radiation risk in perspective, *The Health Physics Society's Newsletter* XXIV 3, 1996). However, in a recent report by Brenner and colleagues it is suggested that there is good evidence of increased cancer risk in humans at 50 mSv although it is unclear from their report what is meant by "good evidence" (Brenner, D.J. et al., Cancer risks attributable to low doses of ionizing radiation: Assessing what we really know, *Proceedings of the National Academy of Sciences,* 100, 13761, 2003). In a report on collective dose, the National Council on Radiation Protection and Measurements (NCRP) reaffirmed the LNT model as an underlying principle of radiation protection but noted that, at low levels, risk estimates

are uncertain by a factor of two or more and a dose threshold cannot be excluded (National Council on Radiation Protection and Measurements [NCRP], *Principles and Application of Collective Dose in Radiation Protection.* NCRP Report 121, National Council on Radiation Protection and Measurements, Bethesda, MD 1995). Canada's Advisory Committee on Radiological Protection has also endorsed LNT theory but recognized significant limitations to its use in predicting health effects below 200 mSv (Advisory Committee on Radiological Protection [ACRP]. *Biological Effects of Low Doses of Radiation at Low Dose Rate*, ACRP-18. (Atomic Energy Control Board of Canada, Ottawa, 1996.) In recent technical reviews of low-dose health effects, the U.S. National Academies reaffirmed the use of LNT theory as a basis for risk estimation and radiation protection practice (*supra* note 11), but the French Academy of Sciences has backed away from support of LNT in the low-dose range. See French Academy of Sciences, National Academy of Medicine, *Dose Effect Relationships and Estimation of the Carcinogenic Effects of Low Doses of Ionizing Radiation*, March 2005, http://www.academie-sciences.fr/publications/rapports/pdf/dose_ effet_07_04_05_gb.pdf (accessed March 2006).

22. Tubiana, M., Dose-effect relationship and estimation of the carcinogenic effects of low doses of ionizing radiation: the joint report of the Academie des Sciences (Paris) and of the Academie Nationale de Medecine, *International Journal of Radiation Oncology Biology and Physics,* 63, 317, 2005.
23. Benninson, D., Sievert lecture: Risk of radiation at low doses, *Proceedings of the 1996 International Congress on Radiation Protection,* 1, 17, 1996.
24. Billen, D., Spontaneous DNA damage and its significance for the "negligible dose" controversy in radiation protection, *Radiation Research,* 124, 242, 1990.
25. Ramsey, M.J. et al., The effects of age and lifestyle factors on the accumulation of cytogenetic damage as measured by chromosome painting, *Mutation Research,* 338, 95, 1995.
26. Whether there is epidemiological evidence of cancer risk at 10 mSv is arguable. See note 21.
27. There are at least two lines of evidence suggesting that repair may be an important determinant of risk. In mammalian cell culture systems, various types of repair phenomena have been well characterized, including subtransformational repair and adaptive responsiveness (Han, A., Hill, C.K., and Elkind, M.M., Repair of cell killing and neoplastic transformation at reduced dose-rates of ^{60}Co γ-rays, *Cancer Research,* 40, 3328, 1980 and Wiencke, J.K. et al., Evidence that the [^{3}H] thymidine-induced adaptive response of human lymphocytes to subsequent doses of x-rays involves the induction of a chromosomal repair mechanism, *Mutagenesis,* 1, 375, 1986). Several human epidemiological studies suggest that low dose-rate reduces risk. See Shore, R.E., Issues and epidemiological evidence regarding radiation-induced thyroid cancer, *Radiation Research,* 131, 98, 1992; Howe, G., Lung cancer mortality between 1950 and 1987 after exposure to fractionated moderate-dose-rate ionizing radiation in the Canadian Fluoroscopy Cohort Study and a comparison with lung cancer mortality in the Atomic Bomb Survivors Study, *Radiation Research,* 142, 295, 1995; Boice, J.D., Risk estimates for radiation exposure, in *Health Effects of Exposure to Low-Level Ionizing Radiation,* Hendee, W.R., and Edwards, F.M., Eds., The Institute of Medical Physics Publishing, Bristol, 237, 1996.
28. Brenner, D.J. et al., *supra* note 21.
29. Anspaugh, L.R., Catlin, R.J., and Goldman, M. The global impact of the Chernobyl reactor accident, *Science,* 242, 1513, 1988.

30. National Research Council, *Health Effects of Exposure to Low Levels of Ionizing Radiation,* BEIR V Report, National Academy Press, Washington, 1990.
31. Brenner, D. J. et al., Estimated risks of radiation-induced fatal cancer from pediatric CT, *American Journal of Roentgenology,* 176, 289, 2001.
32. Ibid.
33. The National Academies BEIR VII Committee recommends further research in this area. See National Research Council, *Health Risks from Exposure to Low Levels of Ionizing Radiation,* BEIR VII Report, National Academies Press, Washington, DC, 2005.
34. Berrington de Gonzalez, A. and Darby, S., Risk of cancer from diagnostic X-rays: estimates for the U.K. and 14 other countries, *Lancet,* 363, 345, 2004.
35. Nuclear weapons function by initiating and sustaining nuclear chain reactions in highly compressed material which can undergo both fission and fusion reactions. Modern nuclear weapons have a trigger called the "pit" that initiates and sustains the explosion. The pit contains plutonium and is a necessary component for nuclear weapons to detonate properly.
36. The draft DOE National Nuclear Security Administration's Environmental Impact Statement for the proposed Modern Pit Facility was issued in May, 2003 http://www.mpfeis.com (accessed March 2006).
37. Pasternak, A., A national solution for a national problem, *Radwaste Solutions,* 28, September/October 2003.
38. National Council on Radiation Protection and Measurements, *Implementation of the Principle of as Low as Reasonably Achievable (ALARA) for Medical and Dental Personnel,* NCRP Report 107, NCRP, Bethesda, MD, 1990; National Research Council, *Health Risks from Exposure to Low Levels of Ionizing Radiation,* BEIR VII Report, National Academies Press, Washington, DC, 2005.

4 Uncertain Risk

Risk assessment is an important tool for informing many kinds of societal decisions. When adverse outcomes (e.g., automobile injuries) are observable directly, risk assessment is usually straightforward and involves few questionable assumptions. However, in most cases risk assessment is a complex process because risk information cannot be measured directly. This necessitates the use of complex modeling involving questionable assumptions. Decisions on cleanup of environmental contamination, decisions to site permanent facilities for the disposal of radioactive waste, and decisions on securing nuclear facilities and materials against terrorist threats require sophisticated risk-assessment calculations. For these applications and others, there is limited experience on which to base estimates of the likelihood and consequences of certain events. The usefulness of risk assessment for decision making is limited by the extent of uncertainty in the analysis. The behavior of complex systems can be difficult to predict because of an imperfect understanding of system parameters (conceptual model uncertainty) or incomplete information about important system properties. Understanding the uncertainties and limitations of risk assessment and conveying those limitations to decision makers in an effective manner remain key challenges for the technical and policy communities.

This chapter explores uncertainties in risk assessment and how they impact subsequent risk-management decisions. Uncertainty does not imply lack of knowledge. Uncertainty is concerned with statistical confidence in data. We have significant information about health effects of radiation at small doses. What we can say is that at doses below about 100 mSv radiogenic cancers are very difficult to measure. Because risks are small, there are large statistical uncertainties in their measurement. Several biologically plausible theories may be used to predict risk at low doses, and the large uncertainties in data at low doses preclude falsification of candidate predictive theories. Risks can take on a wide range of values depending on which theory is used.[1]

For any particular predictive theory, the values of risk coefficients and other theory parameters are important sources of uncertainty. Predictive theories may have multiple parameters, and each parameter may be described by a distribution of values. Sources of parameter uncertainty can be quantified through statistical analysis and include random errors in measurement, systematic errors in measurement, random sampling errors, and use of surrogate data instead of direct measurements.[2] Distinctions between uncertainty and knowledge are not appreciated by the public and result in public misunderstandings and confusion about risks and what we know about them at small doses.

Risk is usually expressed as a numerical probability of an adverse health outcome. This requires that two pieces of information be known about the risk: the nature of the adverse outcome and the probability of its occurrence. For stochastic risks, health consequences (e.g., cancer) are independent of the probability of occurrence. The same outcome occurs at any dose; it is the probability that changes with dose. A solid tumor or leukemia that results from a dose of 100 mSv is the same clinically as the disease resulting from a dose of 1,000 mSv. The probabilistic component of the stochastic risk carries the statistical uncertainties.

Low-dose radiation risk is an example of a risk with uncertain probabilities of well-known consequences. For ionizing radiation for which significant human experience has been accumulated over a wide range of doses, health consequences of exposure are known, but probabilities of occurrence at very small doses are not. Cancer risk associated with medical x-rays is a good case in point. The types of cancer that may occur for different x-ray studies are known, but the probability of their occurrence has a high degree of uncertainty because risk calculations are based on very large dose extrapolations. For some other nuclear technologies, neither the probability nor the consequences are known very well. In the case of permanent disposal of radioactive waste, the long-term environmental consequences and their probabilities are poorly understood. Some risks are well characterized even though the probabilities are small. Obvious examples include lotteries, craps, and other games of chance where the probabilities of winning, the costs, and the prize winnings are known.

Knowledge of risks may be obtained through direct observations of events (usually expressed as a frequency) or by using predictive theories to estimate risks (expressed as a probability). Direct observation is more reliable because frequencies and consequences are determined directly and few assumptions are involved. Frequencies refer to events that have happened in the past. This information (the number of events in a defined reference population) can be used to predict occurrence of events in the future. Generally there is a high level of confidence in such predictions unless event conditions have changed. Probabilities, on the other hand, refer to the chances of a particular event occurring when there is little or no evidence of past occurrences. For very small risks it is practically impossible to distinguish between zero probability and probabilities that are too small to be measured reliably. But absence of evidence of risk is not evidence of absence of risk! Epidemiology is constantly challenged by this zero probability problem.

HOW LOW CAN YOU GO?

We know more about the health effects of ionizing radiation than most other carcinogenic agents.[3] At high doses, radiogenic cancer risks are fairly well known. Risk information has been derived from a number of human epidemiologic studies, most notably the Japanese survivors of the atomic bombings. In the Life Span Study (LSS), which has been going on for more than 50 years, over 85,000 persons have been followed (Table 4.1). Males and females, adults and children, and those exposed prenatally have been evaluated for cancer and other health effects following exposures. Perhaps the most valuable feature of the LSS study is the wide range of radiation doses involved. Individuals were exposed to doses ranging from a few mSv to over 4,000 mSv. No other single epidemiological study offers a dose range covering three orders of magnitude.

TABLE 4.1
Excess Cancer Mortality in Japanese Survivors of the Atomic Bombings at Hiroshima and Nagasaki

Dose Group	Number of Subjects	Excess Solid Cancer Deaths
Controls	38,507	0
<200 mSv	35,909	85
200–500 mSv	6,380	99
500–1,000 mSv	3,426	116
1,000–2,000 mSv	1,764	113
>2000 mSv	625	64

Source: Preston, D.L., What is known about radiation effects at low doses, International Radiation Protection Association International Congress 11, Madrid, May 2004.

Although cancer risks are derived primarily from those survivors exposed to high doses (>500 mSv), in reality the LSS is really a low-dose study. Over 85% of the subjects were exposed to doses below 200 mSv. The number of excess cancers attributable to radiation exposure in the <200 mSv group (the difference in cancer deaths between exposed and unexposed groups) is within random statistical error (Table 4.1). Only at higher doses are the excess cancer deaths large enough relative to the number of subjects exposed to result in statistically significant risk.

The LSS makes a clear statement that at low doses (<200 mSv) radiogenic cancer risk cannot be measured consistently. It is unclear from the data whether the risk is zero or too small to be measured reliably. That risks cannot be measured reliably cannot be used as evidence to support the existence of a threshold. To do so would require clear evidence of absence of risk. The fact that risks cannot be detected at statistically significant levels at doses below 200 mSv does not mitigate the value of the low-dose data.

If no one in the LSS was exposed to doses above 200 mSv, there would be little evidence to support cancer as the major health effects of exposure at low doses. It is interesting to speculate how risk estimates, regulations, and the framework for radiation protection might have evolved without the key epidemiological data above 200 mSv. It is safe to conclude that cancer would have ultimately been identified as an important health outcome at low doses because of experimental studies and also from studies of radiotherapy patients. Although radiotherapy involves very high doses to localized diseased areas, tissues outside the treatment volume do get a small measurable dose that increases cancer risk. Risks of second cancers have been used to corroborate the risk estimates derived from the Japanese Life Span Study.

The ratio of the "signal" (the radiogenic cancer risk) to the "noise" (the natural or spontaneous cancer risk) determines the size of the population needed to detect risk at a given radiation dose (Figure 4.1). Huge populations are needed because the signal-to-noise ratio is very low (only about 2% at 100 mSv).[4]

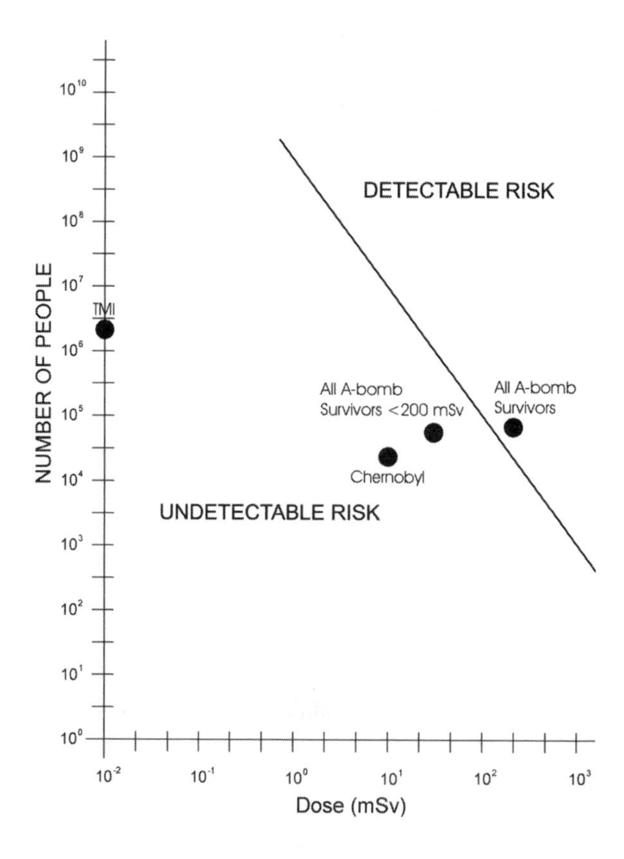

FIGURE 4.1 Population size needed to detect risk. The size of the population necessary to detect a risk is very large and inversely related to the square of the dose.

A population of about 1 billion persons (roughly one-sixth of the world's population) would be required to detect a cancer risk from natural background radiation exposure (assumed to be about 1 mSv per year excluding contributions from radon) if such a risk actually existed. A population of about 10 million would be needed to detect an elevated cancer risk in a population exposed to a dose of 10 mSv (the approximate dose of some medical radiodiagnostic procedures). As a general rule the population size needed to detect a risk with 95% confidence is inversely related to the square of the average population dose. If the average dose to the population is increased tenfold, the population size needed to detect a risk is reduced by a factor of 100.

Several well-documented populations are plotted in Figure 4.1 to illustrate practical difficulties in detecting small risks. The diagonal line establishes the boundary for detection of risk based on population dose and population size. Among the groups shown only the Japanese survivor population was large enough in numbers and had a large enough average dose to detect radiogenic risk. However, risk cannot be detected if only survivors exposed to doses below 200 mSv are considered, even though this represents a sizable proportion for the LSS population (Figure 4.1).

Radiogenic risks are also undetectable in the large Chernobyl recovery worker group and the population surrounding the Three Mile Island (TMI) nuclear power plant.

There was significant public pressure following the TMI accident in 1979 to determine if cancer rates were elevated among the 2 million persons living near the stricken plant. Preliminary calculations clearly indicated that epidemiological studies would be unable to detect an elevated risk even if it were present because the average population dose and the size of the population were too small. Nevertheless epidemiological studies were conducted and to no one's surprise published results showed no radiological causation.[5] About 20% of the exposed population (about 400,000 persons) would be expected to die of cancer in the absence of any radiation exposure. The average exposure to the TMI population was about 0.02 mSv (a dose well within the variations of natural radiation background levels) and two additional cancer deaths due to radiation would be expected theoretically from this dose.[6] Clearly, it is impossible to discern a "signal" (two excess cancer deaths) when the "noise" (400,000 cancer deaths) is so large. The average population dose to the TMI population would need to be about 20 mSv, or 1,000 times higher, to detect a radiogenic risk.

The minimum dose necessary to detect statistically significant increases in radiogenic cancer has been a point of contention in the epidemiology and radiological health and safety communities for years. Evidence of risk at very small doses informs the standards-setting process and can help settle the question of the appropriate predictive theory for use in radiological protection. In children the lowest dose associated with statistically significant radiogenic cancer is 100 to 200 mGy for thyroid cancer based on an analysis of seven pooled studies and in studies of children in Belarus and the Russian Federation exposed to radioiodine after the Chernobyl power plant accident. In adult populations, the lowest dose associated with significant radiogenic risk appears to be 200 mSv based on analysis of the LSS database.[7] However, some have argued that risks can be measured reliably at much lower doses. The principal data set used to support this view is the study of cancers in children irradiated *in utero* in the Oxford Survey of Childhood Cancers (OSCC), which reported a 40% increased risk of childhood leukemia associated with low-dose intrauterine exposure to diagnostic radiation between 10 and 100 mGy. Study of childhood cancers circumvents major epidemiological limitations of studies of adult populations. The developing embryo is considered more radiosensitive than adults, and childhood cancers are relatively rare. Thus the signal-to-noise problem characteristic of studies of adult populations is avoided by increasing the signal (i.e., radiosensitivity) and decreasing the noise (spontaneous cancer incidence in children). However, the results of the OSCC study have not been corroborated in cohort studies of children irradiated *in utero* (principally Japanese women who were at various stages of pregnancy at the time of the bombings), and the findings of approximately equal relative risks for different childhood cancers remain puzzling. Accordingly, the causal nature of an association between *in utero* radiation exposure and childhood cancer and the level of risk remain uncertain.[8] However, even if childhood cancer risks from doses as low as 10 to 50 mSv were observable it is unclear what the relevance would be for cancer risks in adult populations. Studies of thyroid cancer in irradiated populations illustrate that children and adults respond quite differently.

Radiogenic thyroid cancer risk is substantially greater in children exposed early in life than in individuals exposed in later life; adult risks may be 50% or less than the risk in children.[9] Accordingly, more aggressive protective actions are needed for pregnant women and children.

A recent analysis of atomic bomb survivor data from 1950 to 1990 suggested that the dose-response curve for cancer mortality is linear down to 50 mSv and that this is the lowest dose linked to a statistically significant radiogenic risk. But risks cannot be reliably measured below 200 mSv in the Japanese survivors because of population size constraints (Figure 4.1). In other independent analyses of the LSS data, a curvilinear dose response also provided a satisfactory fit to the Japanese data and, using different analytical methods, no evidence for increased tumor rates below 200 mSv was found.[10]

Typical doses from diagnostic medical and occupational radiation exposures are about 100 times smaller than the doses necessary to detect statistically significant radiogenic risks. It would appear that epidemiological-based risks have a minimal uncertainty of about 1% due to errors in ascertainment of disease.[11] This means that risks less than 1% (or 1 in 100) cannot be considered reliable. In the case of the LSS, reliability in estimating risks comes from ascertainment of cancer mortality in subjects exposed at very high doses. At 1000 mSv lifetime cancer mortality risk is about 5%.

In some situations risk uncertainties may be so large that decisions cannot be based on risk and therefore other factors (including economic, political, and social factors) dominate decision making.

RISK ASSESSMENT CONSIDERING UNCERTAINTY

Risk information is not very useful in risk-management decision making without considering the magnitude and sources of uncertainties. If there were no uncertainties, decision making would be easy and straightforward. Selecting specific risk-reduction strategies would simply be a matter of comparing known costs and known outcomes of alternative strategies. However, in reality the effectiveness and costs of risk management options are difficult to assess because risks are not known very well in the dose interval of interest and dose-response relationships are uncertain. The rate of change in risk with dose (i.e., the slope of the dose-response curve) dictates the expected risk reduction for a given diminution in dose. Clearly, costs and public health benefits will depend on the shape and slope of the assumed dose-response relationship. Risks at low doses can differ significantly depending on the predictive theory used to derive them.[12]

Based on Ockham's razor, LNT theory is preferred because it is the simplest one among alternatives.[13] But more powerful theories (containing several parameters) make a wider range of predictions. Single-parameter theories may be more advantageous if the true value falls within the narrow range of predicted values. If not, multiparameter theories may be more desirable because of the wider range of predictions. The problem with multiparameter theories is that the meanings of the parameters may not always be clear. To be useful in risk estimation, all theory parameters should have biological meaning.[14]

Risk assessment usually makes the simplifying assumption that the population at risk is homogeneous with respect to carcinogen sensitivity. But it is well known that children and pregnant women are more sensitive on a unit dose basis to the effects of a broad range of agents, including medicines. Many prescription drugs are contraindicated in pregnancy because of the higher sensitivity of the embryo. There is also recent evidence that a proportion of the population is at higher risk for cancer because of inherited predisposition due to specific germ-line mutations in so-called cancer genes.[15] Substantial uncertainty exists in risk assessment regarding these sensitive subpopulations. Because of population heterogeneity protective actions may apply to certain subpopulations but not to others. In an uncontrolled release of radioactive material from a nuclear power plant decision makers may recommend certain protective actions for sensitive groups such as pregnant women and children located within several miles of the plant. The decision to take protective action hinges on an understanding of the uncertainties in the projected populations doses. Although the central dose estimate may be below the protective action guide (PAG), protective actions may still be advisable if the PAG is within the range of uncertainty in the dose estimates (e.g., 90% confidence interval).[16] Credible dose projections are important because decision makers want to avoid as much dose as possible by taking protective actions, particularly for critical population groups.

Each input parameter in the risk-assessment process also carries uncertainties. Some input parameters may be known with a high degree of confidence because they are based on direct empirical observations; other parameters may be poorly understood because they are derived from shaky theoretical assumptions. The confidence in any risk assessment is only as good as the least precise component in the risk-assessment process. The cancer risk coefficient (often expressed as the number of cancer deaths per 100,000 population per unit dose) is a key input parameter in determining cancer burdens in exposed populations. These estimates may have considerable uncertainty depending on the cancer site and causal agents considered. For tobacco-related lung cancer mortality, the risks are well known because smoking rates and cancer deaths are easily measured. For environmental carcinogen exposures, risk estimates are highly uncertain because cancer deaths cannot be observed directly and individual exposures may be difficult to measure.

Interpretation of very small risks is further complicated by variability in local natural background radiation levels and local spontaneous cancer rates. These sources of local variations must also be considered to put any estimated risk increases (or decreases) in proper perspective. A risk assessment in one geographic location may have a very different meaning from one conducted in a different locale. Natural background radiation levels change according to geographic location and altitude. The background level of radiation can vary by a factor of two depending on the primordial radionuclide content of the Earth's crust locally. Denver and Santa Fe have some of the highest readings in the U.S.; the mid-Atlantic region has some of the lowest.[17] Doses that approximate natural background radiation levels must be interpreted with caution depending on local natural background radiation levels.

Variability in spontaneous cancer rates also complicates interpretation of risks. It is difficult to detect changes in risk following low doses of radiation because spontaneous cancer rates are high and variable. The variability in spontaneous rates may actually be greater than the estimated radiological risk. Spontaneous cancer frequency is related to a large number of variables, including genetic background, environmental carcinogen exposures, cigarette smoke, diet, and lifestyle. Lung cancer mortality rates vary by a factor of 10 across all U.S. counties.[18] Variations in spontaneous cancer rates cannot be explained by geographic variations in natural background rates; other factors such as smoking and diet are major risk determinants.

The U.S. Environmental Protection Agency's (EPA) Federal Guidance Report No. 13 illustrates the difficulties encountered in risk assessment when risk uncertainties are not quantified.[19] This report on health risks from low-level environmental exposure to radionuclides provides radionuclide-specific lifetime radiogenic cancer risk coefficients for the U.S. population based on age-dependent intake, dosimetry, and risk models. Risk coefficients for specific radionuclides are valuable in translating radioactivity to health risks. The report discusses uncertainties in generic terms but tabulates lifetime cancer morbidity and mortality risk coefficients on a per becquerel (Bq) basis for over 100 radionuclides as point estimates without quantifying uncertainties. The implication is that very tiny amounts of radioactivity (1 Bq is equal to one disintegration per second) confers a nonzero health risk.

The lifetime cancer mortality risk of 1.04×10^{-12} per Bq for tritiated water is typical of the magnitude of radionuclide risks discussed in the report. On a per Bq basis, this risk holds little meaning; the reciprocal of the risk is larger than the world's population. Ingestion of several million Bq of tritiated water would result in a lifetime cancer mortality risk of about 1:100,000 to 1:1,000,000. The most probable outcome from ingesting this minute quantity of tritiated water is a zero risk of cancer death.

Risk coefficients presented in the EPA report are based on biokinetic, dose, and risk projection models. Uncertainties in these models have not been fully incorporated in the risk coefficient calculations. Risk coefficients are presented as single numbers reflecting an unwarranted degree of certainty. Risks should be presented as a range of possible outcomes.[20]

A credible risk analysis considers all sources of carcinogen exposure. This is particularly important in workplace environments where exposure to various physical and chemical agents is likely. Interpreting multi-agent exposure data in risk assessment is challenging. Often risks cannot be individually quantified, and it is unclear how individual agents may interact.[21] If agents act independently, risks are additive. Mechanistic interactions of agents complicate matters. Under such circumstances agents may interact in a multiplicative or other complex fashion. In the absence of definitive information, risks are assumed to be additive, recognizing that uncertainties in the total risk may be quite large

UNCERTAIN CHOICES

Risk-management decisions are made to improve public health and safety. Risk-reduction strategies that have little chance of improving public health and safety divert valuable resources from programs that may have significant health impacts. In the occupational setting, risk management is triggered primarily by comparing risks or doses to regulatory limits as set by standards-setting bodies.

Usually several options are available to the decision maker and risk manager. These range from doing nothing to banning the technology altogether and reducing risk to zero (a precautionary approach). The options in between depend on the technology and its perceived benefits; the health risks involved; and economic, social, and political considerations. Decision making requires scientific, social, political, and economic inputs. Even though the risk is below the regulatory limit, should it be reduced further? If so, what are the economic, political, and social consequences of further risk reduction? Is there sufficient scientific evidence to suggest that reducing risk will result in a public health benefit? Should risks be reduced solely because we have the technical means to do so? Risk management choices are discussed in more detail in Chapter 5.

The as low as reasonably achievable (ALARA) philosophy is the cornerstone of radiological risk management. The philosophy centers on keeping doses as low as possible given economic and social constraints. An effective ALARA program results in a residual dose that is generally considered to be acceptable. If the residual dose is not acceptable, additional resources are allocated until an acceptable dose is achieved.

To be credible, risk-management decisions must be based on complete information about the risk, including uncertainty analysis. Scenarios in a typical decision framework in radiation protection are illustrated in Figure 4.2. In radiation protection, a top-down approach is used to control risk using the ALARA philosophy as a basis for protection strategies. Limits establish a regulatory ceiling for risk. Doses are kept as low as reasonably achievable below the ceiling, taking into account social and economic constraints. In many kinds of radiation protection programs, one or more administrative limits or subceilings (set perhaps at 25% or 10% of the regulatory limit) are also employed. Specific protective actions are triggered if doses exceed an administrative limit to ensure that doses remain well below the regulatory limit.

Dose and risk estimates have associated uncertainties that must be considered in comparing values with limits. Regulatory limits and administrative levels do not, by definition, have uncertainties. Highway speed limits may be set at 65 miles per hour but limits are not posted as 65 ± 5 miles per hour.

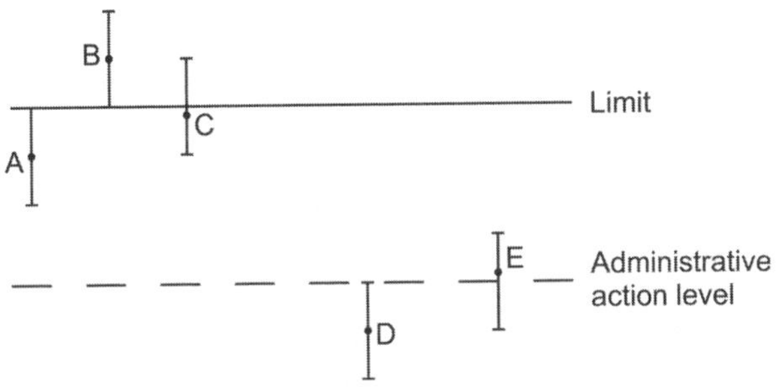

FIGURE 4.2 Decision under uncertainty. Five hypothetical situations are shown where dose or risk estimates and their associated 90% confidence intervals are compared to a regulatory limit and administrative action level. The regulatory limit is a legal limit; the administrative level is set locally.

In situation A the upper bound of the confidence interval is at the regulatory limit. Is this an example of noncompliance? Should protective actions be taken even though the probability of the risk, or dose, is only 5%? In situation D, the upper bound of the confidence interval is at the administrative limit but well below the regulatory limit. Does this situation justify administrative action? For C and E, the situations are less extreme. In these casses, it is more probably than not that the regulatory limit or administrative action level has been exceeded and that protective actions are warranted. What about situation B, where the lower bound of the confidence interval is at the regulatory limit? This is a clear case of noncompliance if the decision is based on the central estimate of the dose or risk. But could a case be made for regulatory compliance based on the lower bound of the confidence interval given that there is only a 5% chance that the value at the boundary is the true dose or risk?

How should confidence intervals be used in decision making? One approach is to base decisions on the upper bound of the 90% confidence interval. This is an attractive approach because it is precautionary and positive risk-management action is taken in the face of uncertainty in risks. The difficulty with this approach is that the calculated upper limit is not very likely to occur and the probability that the upper bound is the true value is only 5%. A lower bound that includes zero should not preclude a risk-management decision from being made.

A less conservative approach is to use the central risk estimate as the basis for decisions because it represents the most probable risk outcome. In this approach compliance would be demonstrated in situations A and D (Figure 4.2). The confidence interval is important as an expression of uncertainty but is not used in decision making. The upper and lower bounds of the confidence interval are not realistic measures of risk and may be unreasonable depending on what assumptions were used in deriving the estimate.

ANOTHER APPROACH

The fact that a risk cannot be reliably measured does not mean that it is unimportant and should not preclude the need to manage the risk. Regulatory limits for carcinogens are set at levels well below measurable levels. There are significant uncertainties in risk at and below regulatory limits. To increase transparency and strive for both public confidence and scientific credibility, full disclosure of uncertainties in risks and agent exposures is needed. Indexing language should be used when reporting "risk" levels to the public and decision makers. Simply reporting a risk as 1:1,000 is not meaningful without some statement about the uncertainty (i.e., margin of error) in the estimate. If there is no context, it is impossible to fully comprehend what the numbers mean. This is similar to news agencies reporting election results (via exit polling) without stating the margin of error. Risk managers depend on uncertainty information to make credible decisions about what, if anything, to do about the risk.

The large uncertainties typically encountered in environmental and occupational settings beg the question of the utility of risk information in decision making. Uncertainties are so large that risk information may not be very helpful in the decision-making process. How can the effectiveness of a particular risk-management strategy be evaluated when diminution in risk cannot be measured? Large uncertainties in risk preclude meaningful decision making. Risk uncertainty can limit rational, fact-based discussions of important questions on decommissioning of existing nuclear facilities, long-term storage facilities for high-level nuclear waste, and construction of new nuclear power plants to reduce dependence on foreign sources of fossil fuels. A shift from a risk-based system to a dose-based system would be more practical and meaningful. Carcinogen doses (or exposures) can be measured at levels far lower than the associated health risks. For many of the important carcinogens regulated by EPA and other federal and state agencies, natural sources of the agent (or sources of closely related chemical compounds) exist for which dose information is available. Natural levels of the carcinogen can be used to set the level of acceptable dose. Epidemiological studies indicate that levels of naturally occurring carcinogens in food and the environment are not associated with health risks. A dose framework eliminates several important sources of uncertainty in risk-based decision making. Without the need to determine risk, uncertainties in dose and cross-species extrapolation and the shape of the dose-response curve are eliminated. The problems presented by a risk-based system of protection and the advantages of converting to a dose-based system are discussed more fully in Chapter 7.

NOTES AND REFERENCES

1. In the 1980 BEIR III report from the National Research Council, three theories were used to predict health effects from exposure to low levels of ionizing radiation. The number of excess cancer deaths predicted under a linear theory was almost 20 times higher than the number predicted by a quadratic (curvilinear) theory. See National Research Council, *The Effects on Populations of Exposure to Low Levels of Ionizing Radiation: 1980*, BEIR III Report, National Academy Press, Washington, DC, 1980.
2. U.S. Environmental Protection Agency, *Guidelines for Carcinogen Risk Assessment*, EPA/630/P-03/001, U.S. EPA, Washington, DC, March 2005. Human variation is also another source of uncertainty and refers to individual differences in biological susceptibility or in exposure. Human variation cannot be reduced but can be better characterized through further research.
3. Substantial information is also available about cigarette smoking and lung cancer. However cigarette smoke contains hundreds of chemicals that may contribute to cancer risks. For the vast majority of chemicals categorized as human carcinogens that EPA regulates little human epidemiologic data are available to estimate risk. Risks are primarily derived from cell and animal studies and extrapolated to human exposure scenarios.
4. National Research Council, Radiation Dose Reconstruction for Epidemiologic Uses, National Academy Press, Washington, DC, 1995; Benninson, D., The Sievert Lecture: Risk of Radiation at Low Doses, *Proceedings of the 1996 International Congress on Radiation Protection*, Vol. 1, 19, 1996; Mossman, K.L. and Marchant, G.E., The precautionary principle and radiation protection, *Risk: Health, Safety & Environment*, 13, 137, 2002; the line in Figure 4.1 is defined by the equation $N = k/D^2$, where N is the population size, D is the average population radiation dose (in mSv), and k is the slope of the line, the value of which is determined by the spontaneous cancer mortality risk in the population (p), the radiogenic risk (r), and the given Type I (α) error. The number of excess radiogenic cancers ($N \times r \times D$) is large enough to be detected with a confidence level of 95% (Type I or α error = 0.05) when

$N \times r \times D > 2 \times \sigma$. The standard deviation, σ, may be approximated by the square root of the total number of cancer deaths.

$$N \times r \times D > 2\sqrt{[N \times p + (N \times p + N \times r \times D)]}$$

Since $N \times p >> N \times r \times D$, then

$$N \times r \times D > 2\sqrt{(2 \times N \times p)}$$

$$N > k/D^2$$

where

$$k = 8 \times p/r^2$$

$$k \sim 10^9 \text{ if } p = 0.2 \quad \text{and} \quad r = 5 \times 10^{-5} \text{ per mSv}$$

(r is assumed to be independent of D under the LNT theory).

5. See Hatch, M.C. et al., Cancer near the Three Mile Island nuclear plant: radiation emissions, *American Journal of Epidemiology,* 132,392, 1990; Hatch, M.C. et al., Cancer rates after the Three Mile Island nuclear accident and proximity of residence to the plant, *American Journal of Public Health,* 81, 719, 1991.
6. Two excess cancer deaths would be expected in a population of 2 million exposed to an average dose of 0.02 mSv assuming a lifetime radiogenic cancer mortality risk of 5×10^{-5} per mSv.
7. Ron, E. et al., Thyroid cancer after exposure to external radiation: A pooled analysis of seven studies, *Radiation Research,* 141, 259, 1995; Shimizu, Y., Kato, H., and Schull, W., Studies of the mortality of A-bomb survivors 9, Mortality, 1950–1985, part 2, Cancer mortality based on the recently revised doses (DS86), *Radiation Research,* 121, 120, 1990; Land, C.E. et al., Early-onset breast cancer in A-bomb survivors, *The Lancet,* 34, 237, 1993; Cardis, E. et al., Risk of thyroid cancer after exposure to ^{131}I in childhood, *Journal of the National Cancer Institute,* 97, 724, 2005.
8. Knox, E.G. et al., Prenatal irradiation and childhood cancer, *Journal of Radiological Protection,* 7, 177, 1987; Brenner, D.J. et al., Cancer risks attributable to low doses of ionizing radiation: Assessing what we really know, *Proceedings of the National Academy of Sciences,* 100, 13761, 2003; International Commission on Radiological Protection, *Low-Dose Extrapolation of Radiation-Related Cancer Risk*, ICRP Publication 99, Annals of the ICRP 35(4), 2005.
9. National Research Council, *Health Effects of Exposure to Low Levels of Ionizing Radiation,* BEIR V Report, National Academy Press Washington, DC, 1990; National Research Council, *The Arctic Aeromedical Laboratory's Thyroid Function Study: A Radiological and Ethical Analysis,* National Academy Press, Washington, DC, 1996; Cardis, E. et al., Risk of childhood cancer after exposure to ^{131}I in childhood, *Journal of the National Cancer Institute,* 97, 724, 2005.
10. Pierce, D.A. et al., Studies of the mortality of atomic bomb survivors, Report 12, Part I. Cancer: 1950–1990, *Radiation Research,* 146, 1, 1996; Little, M.P. and Muirhead, C.R., Evidence for curvilinearity in the cancer incidence dose-response in the Japanese atomic bomb survivors, *International Journal of Radiation Biology,* 70, 83, 1996; Heidenreich, W.F., Paretzke, H.G., and Jacob, P., No evidence for increased tumor rates below 200 mSv in the atomic bomb survivors data, *Radiation and Environmental Biophysics,* 36, 206, 1997.

11. National Council on Radiation Protection and Measurements (NCRP), *Uncertainties in Fatal Cancer Risk Estimates Used in Radiation Protection,* NCRP Report No. 126, NCRP, Bethesda, MD, 1997. Dose and dose rate effectiveness factors, risk transfer between populations, statistical errors, dosimetry, and lifetime risk projections are significant sources of uncertainty in cancer risk estimates derived from epidemiological studiers of the Japanese survivors of the atomic bombings.
12. *Supra* note 1.
13. Ockham's razor, or the Principle of Parsimony, is attributed to William of Ockham, a 14th-century logician and Franciscan friar. Among competing theories the simplest one is preferred, all other things being equal. In science it is used as a loose guiding principle when evaluating competing scientific theories. The theory that contains the least possible number of unproven assumptions is the most likely to be fruitful. By taking the simpler alternative (the one with the fewest parameters), the chance of compounding errors is reduced.
14. Mole, P., Ockham's razor cuts both ways: The uses and abuses of simplicity in scientific theories, *Skeptic*, 10, 40, 2003.
15. Sankaranarayanan, K. and Chakraborty, R., Impact of cancer predisposition and radiosensitivity on the population risk of radiation-induced cancers, *Radiation Research,* 156, 648, 2001.
16. A protective action guide is the projected dose to an individual from an unplanned release of radioactive material for which a specific protective action to reduce or avoid that dose is recommended. Protective actions are interventions (e.g., sheltering or relocation) that must be taken to protect the public during a nuclear event when the source of exposure of the public is not contained or under control.
17. National Council on Radiation Protection and Measurements (NCRP), *Exposure of the Population in the United States and Canada from Natural Background Radiation,* NCRP Report No. 94. NCRP, Bethesda, MD, 1987.
18. See Devesa, S.S. et al., *Atlas of Cancer Mortality in the United States, 1950-94.* U.S. Government Printing Office, Washington, DC, 1999 (NIH Publ No. [NIH] 99-4564). The U.S. National Institutes of Health has evaluated cancer rates on a county-by-county basis. Lung cancer mortality in white males (1970–1994) is 150 per 100,000 in the top 10% of counties and 13 per 100,000 in the bottom 10% of counties. Variation in cancer mortality differs by gender, ethnicity, and types of cancer.
19. Environmental Protection Agency, *Cancer Risk Coefficients for Environmental Exposure to Radionuclides,* Federal Guidance Report No. 13, EPA 402-R-99-001, U.S. Environmental Protection Agency, Washington, DC, September 1999. http://www.epa.gov/radiation/docs/federal/402-r-99-001.pdf (accessed March 2006).
20. The National Research Council's BEIR V Report notes that at doses in the range of natural background levels sufficient uncertainty in risk exists such that the possibility of zero radiogenic risk cannot be excluded. See National Research Council, *Health Effects of Exposure to Low Levels of Ionizing Radiation,* BEIR V Report, National Academy Press, Washington, DC, 1990.
21. Probably more is known about the cancer risks from cigarette smoking and ionizing radiation than any other agents. Yet the nature of their interaction remains elusive. Epidemiological studies of lung cancer in uranium miners who smoke indicate that the interaction of radon gas and cigarette smoking is complex. Risks appear to be nonadditive but whether the interaction is multiplicative, submultiplicative, or some other form of interaction remains to be resolved. See National Research Council, *Health Effects of Exposure to Radon,* BEIR VI Report, National Academy Press, Washington, DC, 1999.

5 Zero or Bust

The goal of risk management is to reduce risk. This can be done by either avoiding exposure altogether or instituting engineering and other controls to reduce dose to workers and the public. In practice it is the dose of the agent that is controlled, not the risk itself. Accordingly, the underlying assumption in risk management is that reduction of dose leads to a concomitant reduction in risk. Unfortunately, there is little direct evidence to support this assumption for carcinogen exposures in occupational and environmental settings.[1] In reality, the number of cancers averted for a given diminution in dose cannot be observed directly because risks are very small to begin with. Instead, the number of cancer deaths averted is calculated based on a theoretically determined reduction in risk.

This chapter explores commonly used strategies to manage risks. Do all risks need to be managed or are some risks so small that they pose little if any health threat and can therefore effectively be ignored? This chapter also explores risk characteristics that trigger management decisions. Whatever management decisions need to be made, they should be based on considerations of dose rather than risk. As discussed in previous chapters, measurements of risk are highly uncertain, particularly at levels typically encountered in environmental and occupational settings. However, we can measure small doses of radiation (particularly from external sources of x- and gamma rays) very accurately. Anchoring dose measurements to natural background radiation levels is a meaningful approach to dose management without incorporating risk estimation and its significant uncertainties.

Management of radiation risk is complicated by fear of radiation and, interestingly, by our technological capacity in measuring very small radiation doses. Radiation can be measured at levels that are tiny fractions of natural background radiation levels. The view is that if we can measure very small doses, then we should be able to control very small doses.

If the assumption that dose and risk reduction are coupled is not tenable, then reducing dose will have little or no public health impact. The shape and slope of the predictive dose-response function determine the amount of risk avoidance that can be expected from a given dose reduction (the types of dose-response functions are discussed in Chapter 2). The shallower the slope, the more dose reduction is needed to achieve a given level of risk reduction.

Dose-risk coupling is complicated by the fact that cancer is a chronic disease that develops years or decades after acute or chronic exposure to the carcinogen. Postexposure events, including injury, exposure to other toxic agents, or the presence of intercurrent diseasme, that lead to death interfere with coupling. If death occurs before the end of the latency period, the cancer that was destined to develop never occurs and the coupling between dose and cancer is lost. In contrast, coupling is almost always conserved for acute effects such as injuries from traffic accidents or exposure to fast-acting toxins because the event or exposure is closely linked in time with the effect. Time is usually so short that it is highly unlikely that some intervening process could interfere with development of the effect as a result of the causal agent. There is usually little question that neurological symptoms seen in a child after a bark scorpion sting are due to anything other than the scorpion venom.

A shift to a dose-based system of management avoids the dose-risk coupling problem. Because natural background radiation is so well characterized[2] dose reduction can be put into meaningful context through appropriate natural background comparisons.[2] Furthermore, as discussed later in this chapter, the underlying framework in radiation protection is predicated on maintaining radiation doses below regulatory dose limits using an as low as reasonably achievable (ALARA) philosophy. Such an approach to protection is strictly dose-based and does not require calculations of risk.

There are four basic approaches to risk management: (1) eliminating or avoiding the risk entirely, (2) avoiding unacceptable risks, (3) avoiding unacceptable costs, and (4) balancing costs and benefits.[3] The only way to achieve zero risk (i.e., total safety) is to avoid exposure entirely. This means abandoning the agent or activity. Application of the precautionary principle (discussed in detail later in this chapter) often involves abandoning or eliminating a product or activity in favor of safety. Such avoidance measures can have serious risk trade-off consequences as illustrated by the impact of the discontinued use of dichloro-diphenyl-trichloroethane (DDT) as a pesticide for the control of disease-carrying mosquitoes. Alarming increases in malaria have been documented in Third World countries and other regions where DDT use has been terminated.

Avoiding unacceptable risks requires that risks be reduced to tolerable levels based on derived benefits (e.g., reduction in cancer). There is no requirement to reduce risks to zero (i.e., maximize benefits), but the goal is to achieve a desirable level of health benefits. In setting standards for hazardous air pollutants, the U.S. Environmental Protection Agency (EPA) has defined acceptable risk as an individual mortality risk of no greater than 1:10,000. Although the approach is targeted to achieve a particular level of risk, there is no consideration for the costs necessary to meet the risk target. Under the acceptable-risk approach, significant compliance costs can result with little public health improvement.[4]

Avoiding unacceptable costs is the flip side of the acceptable-risk approach and focuses on the costs of regulations rather than the benefits of risk reduction. In this approach, risks are reduced as much as possible while keeping compliance costs below an unacceptable level. This approach disregards benefits of risk reduction. If a risk-management program with exceedingly high costs would also save many lives, society would be better off with the program even though it is expensive. But, under the acceptable-cost approach, opportunities to achieve significant public health gains may be rejected because compliance costs are too high.[5]

The final approach combines the acceptable-risk and acceptable-cost approaches and seeks to provide a balance of costs and benefits. By considering costs and benefits together, net benefits can be optimized. The optimization principle is a cornerstone of radiation protection. Once a radiological practice has been justified, risk-management decisions must consider how best to use available resources to reduce radiological risks to individuals and the population. An ALARA approach is used to reduce doses where the amount of dose reduction is constrained by economic and social factors. Any residual risk as a consequence of an ALARA program would be considered acceptable (otherwise, additional resources would be allocated to reduce dose further) and protection would then be considered optimized.[6]

Although health and safety regulations are directed primarily toward businesses that manufacture products or provide services that subject workers or the public to health risks, individuals also have a responsibility to manage risks on a personal level. Risk management is everyone's business whether one is talking about managing risks of sophisticated technological products or making good personal hygiene choices. Brushing one's teeth is a simple risk-management activity to reduce the risk of tooth decay. The decision to brush is not based on sophisticated calculations of costs and benefits. Instead, individuals rely on knowledge and experience, advice from experts (e.g., dentists and dental hygienists), and intuitive balancing of benefits (e.g., good oral hygiene, a healthy smile) against costs (e.g., expensive dental work). In general, people choose which risks to avoid or to minimize, and which ones to accept by balancing benefits of activities or products against risks. However, people are not necessarily rational. Some activities such as smoking and chronic alcohol consumption carry risks that are far larger than the benefits. People choose to engage in these behaviors anyway because the *perceived* risks are considered to be outweighed by the *perceived* benefits. Risk management at the individual level is a matter of perceptions and control of individual behaviors. It should be noted that managing individual risks can be influenced by social factors out of control of the individual. Limited resources, because of socioeconomic factors, may limit choices and thus impact the individual's ability to alter behaviors. Obese individuals living and shopping in poor neighborhoods may have limited food choices, therefore restricting ability to control food intake and thus control weight. Unlike technological and industrial activities and products that are controlled through regulation, personal behaviors cannot be easily controlled through government intervention.

Individuals at risk may not necessarily reap the benefits, but when the individual at risk also benefits, efforts to manage dose (and risk) often will result in increased benefit. This is encountered in diagnostic medical imaging where efforts to reduce patient dose can increase diagnostic quality of the study.[7] Patients do not benefit when diagnostic studies need repeating because diagnostic images are poor quality. Repeat studies increase patient overall dose. The need for repeat studies is an ongoing problem in diagnostic radiology.

MANAGEMENT TRIGGERS

Risk management is a complex decision process involving policy alternatives and selecting the most appropriate regulatory action. Risk management is not science in the way risk assessment is (i.e., measurements of exposure, dose response, uncertainty analyses, etc.). Instead, it combines quantitative and qualitative information about risk with economic, ethical, legal, political, and social judgments to reach decisions.

There are two major decisions to be made in risk management: (1) which risks require management and which risks can be ignored? and (2) if the decision is to manage, what methods or processes should be used to reduce risk and how much of the risk should be averted? Not all risks require management; some naturally occurring risks may be significant but cannot be adequately managed because they cannot be controlled (e.g., tornadoes and earthquakes).

What factors are important in triggering a risk-management decision? The most obvious concerns relate to technical aspects of the risk, such as the magnitude of the risk and the size of the population affected. Large risks regardless of the population size at risk must be managed if they can be controlled. For small risks, how the risk is physically distributed in space and in time may be more important than the magnitude of the risk. Social and political factors are also of considerable importance since they are concerned with perceptions, feelings, and values, and how risks and benefits are distributed. Political and social valuations of the risk are not quantifiable but involve issues of political power and fairness (i.e., distributive justice, procedural justice). Closely related to these social characteristics is society's prioritization of risks. As discussed in Chapter 6, society's ranking of risks does not coincide very well with technical attributes such as actual probability or consequence of the risks.

The challenges often encountered in risk-management decisions are embodied in the conflicting technical and social values in risk analysis. Although current models of risk analysis (e.g., the National Academy of Sciences' Red Book) do not explicitly incorporate non-scientific factors, value judgments do play a central role (e.g., selecting a particular dose-response function is a value judgment). Experts usually subscribe to technical models of risk analysis and have difficulty understanding or appreciating the views of lay persons who discount quantitative measures in favor of qualitative aspects as developed in social models of risk analysis. Similarly, lay persons cannot understand risk-management decisions that are based on strictly quantitative approaches to risk analysis. The technical model is characterized by values of rationality (risk analysis is a quantitative, logical, and consistent process), objectivity (even though a risk may be present, it is not relevant if it cannot be measured), efficiency (risk management should make the best use of available resources), and meritocracy (scientists and technical experts should make decisions because they are knowledgeable and know best). The social model holds very different values and includes subjectivity (perceptions and feelings about risks are as or more important than quantitative measures), experience (theory and calculations are not as important as history in evaluating risks), socioculturalism (fairness as reflected in distributive and procedural justice, and democratic values), and pluralism (pluralistic society offers a collective wisdom beyond what the expert community can achieve).[8]

Technical Triggers

The quantitative elements of the risk-assessment process form the basis for technical triggers for risk-management decisions. Hazard assessment, exposure assessment, dose-response assessment, and risk characterization are discussed in some detail in Chapter 1.

Size Matters

The size of the risk and its distribution in the population are key management triggers.[9] There is little public argument that large risks to a few individuals or to a large population should be controlled to the largest extent possible. Large risks are readily observable and fairly easy to measure, and there is usually little question that the agent or activity in question was the causal factor. For example, air travel would be very risky if there were no safety regulations to ensure the proper functioning of equipment and constant training of pilots. Local traffic laws provide controls on alcohol consumption. Without such controls the numbers of alcohol-related traffic injuries and fatalities would skyrocket. Although these regulations do not eliminate accidents entirely, they provide a level of control that results in an acceptable level of risk. Legislative action can always make regulations more restrictive if society deems the residual risks to be unacceptable.

The situation with small risks is somewhat more complicated. Management decisions hinge, in part, on the nature of the risk and the population at risk. Is exposure to the agent or activity likely to result in permanent injury or transient health effects? Unlike risks from automobiles or airline travel that can be measured directly, cancer risks have large uncertainties. Regulatory limits for carcinogens (including ionizing radiation) are set at levels far below observable health effects in order to establish a large safety margin.

Radiation doses to patients undergoing diagnostic procedures are not subject to regulatory limits established for workers or the general public. Pediatric computerized tomography (CT) or cardiac angiography examinations may involve substantial doses to the patient. Doses from a single study can exceed allowable annual doses to nuclear workers. Nevertheless, clinicians have become increasingly aware of the need to control doses especially in interventional procedures. As a risk manager, the clinician must use professional judgment to balance risk and benefit. The clinician seeks to minimize patient dose (reduced number of films, limit repeat studies, limit fluoroscopy time) while ensuring that the diagnostic quality of the study is not compromised. A study that requires repeating because of poor diagnostic quality due to efforts to reduce patient dose is an unacceptable practice.

The EPA uses a risk system as a common currency for the large number of carcinogens it regulates. It is assumed that the risks from exposure to various cancer-causing agents are comparable and can be combined. For environmental cleanup at Superfund sites, risk-management decisions are primarily driven by legal requirements. The EPA is responsible for developing risk-assessment guidelines under the Comprehensive Environmental Response, Compensation and Liability Act (CERCLA), also known as Superfund. Under CERCLA, excess lifetime risks of cancer for carcinogenic contaminants in the range of 10^{-6} to 10^{-4} are considered to be acceptable. To put this risk range into perspective, the cancer incidence in the U.S. from all sources is about 1 in 3. An excess lifetime risk of cancer is the probability above the 1 in 3 risk of developing cancer in the U.S. If the excess cancer risk is above 1 in 10,000, then an action must be taken to reduce the cancer risk from exposure to this contaminant.[10]

Risk-management practices to control small cancer risks hinge on the assumption that dose reduction has a public health benefit. The slope of the dose-response curve (a quantitative measure of the risk per unit dose) determines the amount of risk reduction for a given diminution in dose. In most instances, the slope of the dose response (assumed to be constant over the dose range of interest) is not known very well, and it is a matter of faith that dose-reduction practices actually have a public health benefit.

Sensitive People

Several genetic diseases confer heightened cancer risks and have raised questions about whether a utilitarian approach to risk management provides adequate protection of sensitive subpopulations.[11] It is generally assumed that carcinogenic responses to ionizing radiation and chemical carcinogens follow a Gaussian or normal frequency distribution (Figure 5.1). Some individuals will show a greater sensitivity to a given dose of the agent; others will be more resistant. But recent studies in molecular genetics suggest that a random distribution of radiosensitivity may be overly simplistic. Much of human variation in sensitivity to environmental agents is now thought to result from individual gene variations due to single nucleotide polymorphisms (SNPs), specific gene codon variations that result in a change of a single amino acid inserted at a specific site in the protein product of the gene. SNPs may be an important determinant of individual radiosensitivity. SNP variations are not normally distributed but appear to be concentrated in certain regional and ethnic groups. Ethnic and regional differences create serious policy questions regarding

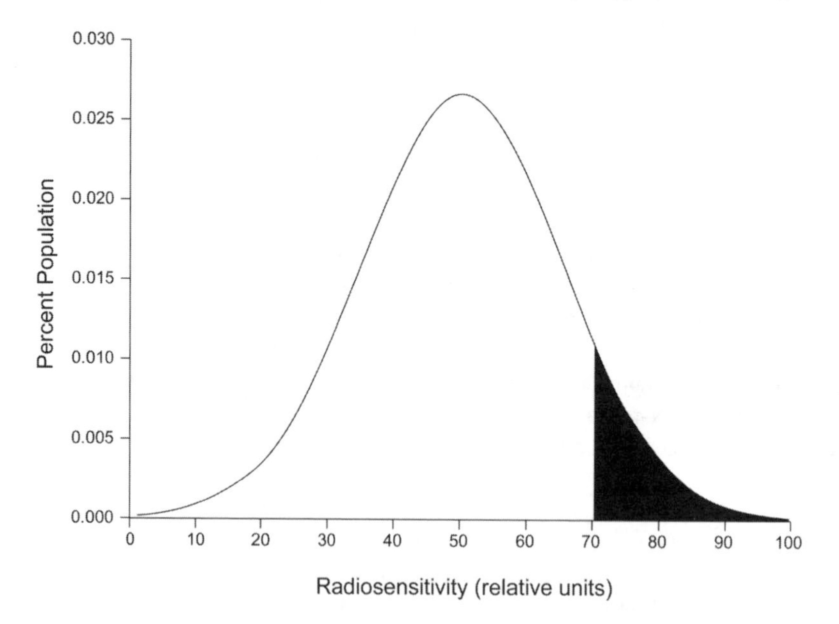

FIGURE 5.1 Theoretical distribution of radiation response in a human population. The darkened area is 10% of the total area under the curve and represents a theoretical subpopulation with heightened radiation sensitivity. Dose limits based on the typical or average radiation sensitivity may not adequately protect subpopulations that are highly radiosensitive.

protection of specific individuals and population subgroups. Studies of cancer genes and SNPs have emerged as important research needs to elucidate the underlying basis of variations in individual sensitivities to cancer.[12]

The possibility that children or pregnant women may be at risk may trigger additional or special risk-management actions. In general, risks from carcinogen exposures are higher in children because they are expected to live beyond the disease latent period (usually 10 years or more) and express the disease. Children may also be more sensitive because critical tissues such as the bone marrow and central nervous system are still developing or growing. The embryo and fetus are more sensitive to the effects of carcinogenic agents because of the rapid cell proliferation and development that are going on prior to birth.[13] Children undergoing diagnostic x-ray procedures may be at greater risk for cancer. Certain types of studies such as CT examinations can deliver unusually high doses. The radiology community in conjunction with machine manufacturers is cognizant of the need to reduce patient risk through appropriate dose-reduction strategies.

An individual's sensitivity to ionizing radiation exposure has emerged as an important consideration in protection of workers and members of the public. Identifying radiation-sensitive individuals (through medical screening or disease diagnosis) and providing adequate protection raise important policy questions. Should worker and population dose limits be made more restrictive to account for sensitive individuals? Should workers be treated differently because of increased radiation sensitivity? What are the social and economic costs associated with identifying sensitive individuals and providing additional protective measures?

The percentage of workers and the public who have increased risk of radiogenic cancers because of genetic susceptibility to cancer is not known but has been estimated to be in the range of 1% to 10%. This estimate is highly uncertain because it is based on limited epidemiologic and scientific data. Even if 10% of the population is radiosensitive, there is probably little justification to alter current radiation protection limits or practices.[14] Further, a definition of radiosensitivity for the purposes of radiation protection has not been clearly established; the number of radiosensitive individuals in the population will depend on how radiosensitivity is defined.

Risk management of radiosensitive groups presents challenging problems. Do we treat all radiosensitive individuals in the same way, or do we recognize that some health outcomes are more serious than others? Individuals who are at increased risk for nonmelanoma skin cancers are likely to require different risk-management strategies than individuals at increased risk for colon cancer or lung cancer. Although age, gender, smoking, and diet are the principal determinants of cancer risk, individual radiosensitivity may also be important in some circumstances. Radiosensitive individuals working in radiological environments may be at higher risk for cancer. It may be prudent to identify such individuals and provide additional protective measures.

If it is decided that radiosensitive individuals should have enhanced protection, what risk-management strategy should be adopted? The whole population benefits by setting dose limits to account for the most sensitive subpopulation. However, monetary and other costs needed to comply with the lower limits may be substantial and offset the increased public health benefit. Should radiosensitive groups be stratified in accordance with the severity of the cancer predisposition? The current radiation protection framework includes special considerations for pregnant workers. These workers are subject to more restrictive dose limits during pregnancy because of increased embryo/fetus radiosensitivity.

It should be noted that there is considerable interest in identifying sensitive individuals who are candidates for cancer therapy. Screening cancer patients for radiation (and chemotherapy) sensitivity could be useful in identifying optimum treatments. For example, if a patient with prostate cancer has the option of surgery or radiotherapy treatment, information about radiation sensitivity would be important in the treatment decision.

Assigned Blame

The assigned share of risk (also referred to as the probability of causation)[15] recognizes that cancer (and many other diseases) has multiple etiologies. Assigned share (AS) quantifies that portion of the total risk to the agent in question.[15] In principle, AS should be an important consideration in risk-management decisions but is rarely used for that purpose. If a particular agent has a low AS (i.e., agents other than the one in question are more important causes of disease), little public health benefit is gained by assigning significant resources to manage the agent. Causal agents that are both necessary and sufficient (e.g., HIV as a cause of AIDS) will have an AS of 100% because only the agent causes the disease and if the individual has the disease, he or she was exposed to the agent at some time before the onset of disease. The disease does not develop without prior exposure to the agent.

Agents that cause cancer have AS values that range from 1% to 99% depending on the cancer type, dose of the agent in question, doses of other known causal agents, age at exposure, age at diagnosis, genetic predisposition, and other factors that contribute to baseline cancer risk. Evaluating the public health impact of radon must take into consideration the role of cigarette smoking as the major cause of the disease. For smokers living in a typical U.S. home, the AS of radon as a cause of lung cancer is 10% or less.[16] Radon is a minor contributor to lung cancer mortality.

Leukemia can be caused by ionizing radiation and benzene. Only one of the factors need be present to cause the disease. A worker exposed to both benzene and ionizing radiation may get leukemia, but it is not possible to know which agent caused the disease. The retrospective AS calculation estimates the probability that one of the agents was the cause based on dose histories for each of the agents.

AS is used currently as a retrospective decision tool to adjudicate compensation claims in cases of occupational cancer. As provided by legislation, claimants would be entitled to monetary awards if the AS from exposure to the agent in question exceeded a threshold percentage. Under the Energy Employees Occupational Illness Compensation Program Act of 2000 (EEOICPA), claimants are entitled to a lump sum payment of $150,000 plus medical benefits as compensation for cancer resulting from radiation exposure or chronic beryllium disease, or silicosis. Eligible employees must have incurred their exposures while in the performance of duty for the U.S. Department of Energy (DOE) and certain DOE vendors, contractors, and subcontractors. An eligible worker seeking compensation for cancer is eligible for compensation only if the cancer was "at least as likely as not" (an AS value of 50% or greater) caused by radiation doses incurred in the performance of his/her official duties. Calculations of AS including appropriate uncertainty analyses assist the government in determining who is eligible for compensation under the Act.[17]

AS values should be interpreted with caution. Expressing AS as a single number suggests a degree of certainty not supported by science or epidemiology. Point estimates should be accompanied by confidence intervals that reflect the degree of statistical uncertainties in the calculation. The input parameters in the AS calculation are based on population-derived risk estimates. AS is the property of a specified population and not of an individual in that population. AS calculations are not probabilities in the usual sense and are properties of the group to which a person belongs, but in practice are assigned to the individual. It has also been argued that AS is a logically flawed concept, subject to substantial bias and therefore unsuitable. The requirement to use AS retrospectively (i.e., the individual has a documented exposure preceding a cancer that has already occurred) opens the door to serious misuse. Prospective calculations of AS where the individual has not developed cancer can lead to confusion and fear. Basing a retrospective AS computation on the assumption that a person will contract cancer at some undefined time in the future without disclosing the minimal likelihood of the disease occurring is an inappropriate use of the AS calculation. In such cases, AS may be incorrectly interpreted as the actual probability that the individual will get the disease in the future.[18]

Prospective AS calculations have utility in risk-management decisions as long as limitations are accounted for. Knowledge of the contribution of a particular agent relative to other factors in causing disease in a population at risk is valuable in deciding risk-management strategies and in allocating limited resources. Statistical uncertainty in AS is a limiting factor in decision making since risk estimates are input parameters. As discussed throughout this book, estimates of risks from small doses of carcinogens are highly uncertain. Consideration should be given to revising the AS formulation to avoid using risk estimates as input parameters. One approach would involve comparing doses of agents rather than their risks. In the case of ionizing radiation, natural background radiation levels have been measured accurately and can be used as a basis for comparison with other occupational or environmental radiation doses. The necessary assumption is that doses from all sources of radiation are known and the risk per unit dose from different sources is the same. This idea of comparing occupational or environmental doses to doses from natural sources is developed more completely in Chapter 7. However, such an approach is limited here because AS is evaluated only in the context of radiological sources; nonradiological factors (including diet and smoking) dominate cancer causation and would not be included.

Social Triggers

The cultural, economic, political, and social values attributed to an activity or product are also triggers for risk-management decision making. The importance of social triggers is reflected in the differences in the way experts and the public rank order risks. Experts typically rely on quantitative characteristics, including number of expected deaths and population characteristics. The public takes a broader perspective and ranks risks according to personal experiences, feelings, and perceptions. The result is rank order lists that have little in common with one another.[19] The regulatory agenda is more in line with public priorities than with the expert ones. Prioritizing risks is the subject of Chapter 6.

Values underlying social triggers may be procedural or substantive. Procedural values are those that pertain to fairness, justice, transparency and stakeholder involvement in the decision-making process. The public wants a fair, democratic decision-making process that gives balanced consideration to technical and nontechnical issues and provides for a process whereby interested and impacted stakeholders can have meaningful input into decisions. Although scientific and technical perspectives are important in decision making, the process is unbalanced, biased, and incomplete without consideration of socioeconomic, cultural, and political concerns. Substantive values are concerned with decision outcomes, including issues of safety, and protection of children and the unborn.

Safety

Safety is the prime risk-management trigger. There is no substitute for safety, and it is consequently at the top of everyone's list of concerns. Any threat to safety, whether real or perceived, triggers risk-management decision making to reduce or eliminate the risk. But what is "safe"? Safety can be defined as a judgment of the acceptability of risk. Accordingly, a product or activity is safe if its risks are judged to be acceptable.[20] Since "safety" is a question of judgment, what is considered "safe" by one individual or group may not be considered so by others.

Most individuals would agree that a product or activity is safe if it bears no significant additional risk to life and health. For example, societal risks to life and health from nuclear power plant operations should be comparable to or less than the risks due to electric generation from coal, gas, or hydroelectric plants.

There are two approaches to achieving absolute (100%) safety or zero risk: (1) eliminate the product or activity altogether and (2) reduce the risk to zero by using appropriate engineering controls and other risk-management resources. In the first approach, actions to eliminate one risk by abandoning a product or activity may introduce unintended countervailing risks either directly or indirectly. The precautionary principle (discussed later in this chapter) advocates this approach when there are unresolved questions about safety. Unintended consequences of risk management are also discussed later in this chapter. In the early 1990s an epidemic of cholera in Peru claimed over 10,000 persons and infected about 1 million. The epidemic was brought about in part because of the Peruvian government's decision to stop chlorinating drinking water supplies based on the belief that water chlorination would increase cancer risk. Had the water been properly treated, the cholera outbreak would probably not have occurred.[21] The discontinued use of DDT eliminated a potential cancer risk from pesticide exposure, but the policy had other serious public health consequences. Third World countries in particular suffered dramatic increases in malaria as a consequence of the DDT ban. According to the World Health Organization, malaria kills about one million people annually, and there are approximately

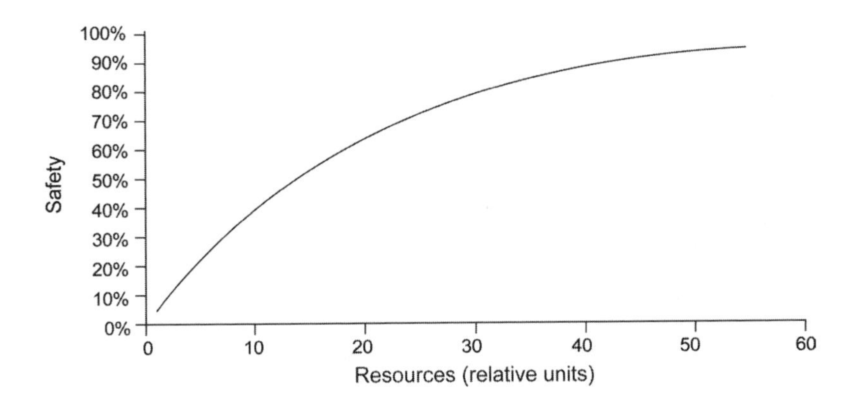

FIGURE 5.2. Theoretical safety performance curve. Improvement in safety (defined as 1-risk) is a nonlinear function of resource allocation. Regardless of the amount of available resources, 100% safety (i.e., zero risk) is never achieved.

300 million to 500 million new cases diagnosed annually. The ban on DDT is not the sole reason for this public health tragedy, but it is clearly a major determinant.[22]

In the second approach, the activity or product is not abandoned but resources are allocated to reduce risks. As shown in Figure 5.2, the greatest safety gains are achieved when initial risks are large. When risks are small to begin with, further reductions become increasingly difficult to achieve, and increasingly more resources must be applied. Zero risk or complete safety is never achievable because human error and system design and engineering flaws cannot be entirely eliminated. There is always some residual risk regardless of the amount of resources allocated to improve safety.

Protection of Children and the Unborn

Concerns for health and safety are elevated by threats to children and the unborn. There is a biological rationale for this because of the increased sensitivity of children and fetuses. But even more importantly there are well-established norms in almost every social and cultural setting that place the highest priority on protecting women and children from harm. Such norms have guided individual and group behaviors to protect pregnant women and children because they represent the link from one generation to the next, and the future and continuity of the social and cultural group. Historical accounts of disasters such as the sinking of the *Titanic* describe rescuing women and children first. Following the Three Mile Island accident near Harrisburg, PA, in 1979, Governor Richard Thornburg issued an evacuation advisory to all pregnant women and preschool children living within a 5-mile radius of the plant as a safety precaution. The accident occurred on March 26 and by March 31 about 250,000 residents (including pregnant women and children) had evacuated the area. The evacuation advisory was lifted on April 9 and families returned to their homes.[23]

Polluters Should Pay

Companies that pollute the environment or endanger public health and safety in the process of producing and distributing goods and services should be responsible for their actions. If technology creates a public health or environmental problem, then technology should fix it. Having someone to blame is a powerful trigger for risk management because offending institutions can be easily targeted for regulatory control and economic relief.

Catastrophe and Apathy

Many types of natural risks, including earthquakes, tornadoes, and tsunamis cannot be prevented or adequately controlled. The public generally accepts these events as "acts of God" since there is no one to blame for the disasters. Although potentially catastrophic natural events cannot be prevented, the public has the expectation that local, state, and federal government will take whatever actions are feasible and necessary to provide early warning and disaster assistance. In the days following Hurricane Katrina on August 29, 2005, inhabitants of New Orleans and other Gulf Coast other areas devastated by the disaster expressed outrage at the inadequate government response. If naturally occurring risks have little catastrophic potential and are perceived by the public to be of little consequence, there may be substantial public apathy because there is no one to blame. Widespread apathy can have a dampening effect on response to government recommendations and advisories. Although risk analysis may suggest that an agent has significant public health potential, the public may consider the government message to be unconvincing, or may consider the public health threat as relatively unimportant.

Measurements of radon concentrations in homes and remediation to reduce indoor radon levels have met with widespread public apathy. The EPA considers radon to be a very serious health risk. Radon tests are easy to perform and are inexpensive. But the public has been reluctant to do anything about the problem because health risks are not considered serious and there is no one to blame for the problem.[24] To address public apathy, the EPA resorted to aggressive campaigning to urge public action. Billboards urged citizens to call 1-800-RADON because "radon is a health hazard in your home." The EPA even ran an Ad Council TV spot showing children playing in a living room. When exposed to radon gas, they suddenly turned into skeletons. But the campaign backfired because the public protested the outrageous and fear-mongering messages.

Public Information and Distorting Risks

The public gets most of its information about risks from print and broadcast media. Risk perceptions are influenced by what people hear, read, and see. Responsible journalists try to provide balanced reports on controversial issues, including whether low-level carcinogen exposures are really harmful. However, the motivation to sell newspapers and to increase numbers of television viewers and radio listeners results in sensationalized stories and reports that tend to distort health hazards. The public is continuously bombarded with reports about smaller and smaller health risks. The public believes that small technological risks are a public health problem and want something done about them. The media focus on small risks has made it difficult for the public to distinguish between theoretical risks and real risks that are based on hard evidence. The result is that the public demands allocation of limited resources (through legislative and regulatory action) to control risks that have little impact on the public health.[25]

A three-article series on the hazards of electromagnetic fields published in *The New Yorker* raised serious health concerns about power lines, cell phones, and computer terminals.[26] At the time these articles appeared in June 1989, video display terminals were common in offices and many homes, and cell phones were becoming popular particularly in the business world. *The New Yorker* articles were based initially on epidemiological findings that among children dying of leukemia in Denver, Colorado, an abnormally high percentage lived in proximity to power lines. The publicity served to produce support for research by state and federal government agencies, individual utilities, and industry organizations. Numerous reviews and analyses of data by government committees and advisory groups concerning health effects of electromagnetic fields were unable to substantiate health claims made in *The New Yorker* articles. Instead they concluded that if a health risk exists from low-frequency electromagnetic fields, it is too small to be reliably measured.[27] As discussed in the case study on cell phones (Chapter 9), managing technological risks can get out of hand. Misguided interpretations of scientific data can lead to imprudent public policy.

Political Triggers

Most risks that concern the public are technology derived (e.g., air pollution, nuclear power, cell phones, etc.). Even though many of these health risks are small and have little impact on the public health or the environment compared to larger risks such as cigarette smoking or automobile travel, they demand attention because of political pressures or social concerns. There is little public tolerance for risks that are derived from technologies.

Public involvement in risk management not only makes sense but is critical to the success of any risk-management effort. The public cannot be expected to accept technological risks if it does not have some say about how the risks are managed. Risk management should involve government (as policy makers and regulators), industries (that generate products and services that subject the public to risks), and private citizens and groups (that are interested in or impacted by the risks). Social institutions have important perspectives and should understand at least at a minimum level the nature of the risks that they voluntarily take or that are forced upon them. Similarly, scientific and technical experts should be aware of the social, economic, and political impacts of risk analysis. A strictly quantitative analysis can exclude important social perspectives. Decision-making inputs regarding technologic risks should not be left to the sole discretion of scientists, engineers, and other technical experts. Although scientific discoveries and advancements are valued, the public also recognizes that these expert groups may have biases. Experts who promote the medical benefits of radiation may downplay radiation risks.

Input from the various stakeholder groups is critical to the development and implementation of effective and balanced risk-management strategies. One real challenge in the public policy and risk-management strategy process is determining to what extent public groups should participate, and when in the process should public participation take place. Although having no public involvement makes it easier for decision makers in the short term, implementation of policies and decisions is made very difficult in the long term because important stakeholder groups are excluded from the decision process. Full participation by all constituents, on the other hand, may slow the process to a crawl, making it difficult, if not impossible, to develop and implement policies in a timely fashion. Unless effective leadership is in place, quality decision making may be compromised by personal agendas, reliance on selected scientific studies, and limited perspectives on the issues.

Stakeholder participation involving interested and impacted individuals and groups is critical in establishing respectability and credibility in risk management and standards-setting processes. Authoritative bodies and the standards-setting organizations they advise should recognize that decision making must be honest, open, and transparent. The integrity and credibility of decision making depends on appropriate public participation. A strictly expert-driven approach to standards setting and risk management eliminates pluralistic views. Experts may have critical technical capabilities not shared by the public, but experts are not infallible and may not be the most legitimate judges of values in risk-management decision making. Unilateral decision making by experts may lead to a public sense of unfairness. Experts, like everyone else, are subject to errors calling for checks and balances by other groups, including stakeholders. Legal actions by public interest groups against DOE and its contractors concerning operations at the Fernald Plant and the Hanford facility led to reassessing of government radiation releases and dose estimates to offsite populations. Public pressure resulted in the U.S. Centers for Disease Control and Prevention to fund and conduct dose reconstruction studies at those facilities.[28]

National and regional policy differences are instructive in identifying key social and political drivers of health and environmental risk management. National wealth is a key driver. Third World countries do not have the economic resources to manage key public health risks. Infectious diseases continue to dominate public health concerns in Third World countries but are essentially controlled in developed, industrialized countries. Political systems also shape national and regional public health and environmental agendas. Decentralized government can lead to nonuniform decision making as reflected by regional politics and priorities. The Nazi national health program to reduce cigarette consumption and in other ways promote good health is an example of how a political dictatorship can effectively implement a national public health agenda.[29]

The case studies on indoor radon (Chapter 8) and cell phones (Chapter 9) explore sociopolitical factors such as public health priorities and philosophies (e.g., the precautionary principle) in shaping risk-management agendas.

Perceptions and Conflicts of Interest

Individuals constantly make risk-management decisions to protect their health and safety. Management triggers are governed by personal experiences and perceptions. The clearest example of personal risk management pertains to decisions about health care.

An individual faced with several treatment options relies on advice from experts (clinicians), acquired knowledge, and input from family and friends. The idea of getting a "second opinion" is not to question the judgment of one expert but to obtain a diverse perspective so that the decision maker can come to an informed decision.

Personal judgments regarding risk management can be influenced by professional or financial conflicts of interest. Industry scientists and technical experts judge environmental risks as less serious than do government scientists or academics.[30] Conflicts of interest in medical therapy decisions may have serious consequences for the patient. Published views and opinions of clinicians and scientists on the safety and efficacy of drugs and food additives have been found to be strongly correlated with their financial position with manufacturers.[31] Clinical research sponsored by drug companies also impacts the practice of medicine. There is compelling evidence that funding of drug studies by the pharmaceutical industry is associated with results favorable to the research sponsor.[32] Clearly professional or financial conflicts do not always lead to ill-advised risk-management decisions, but the potential influence of such conflicts on personal judgment, particularly in medical care, is troublesome. Whether real or perceived, conflict of interest erodes confidence in decision making. Viable options may be discounted or ignored because of conscious or subconscious bias.

Risk-management decisions are challenging because the issues and concerns that experts share are different from those of the public at large. Expert assessments and social values can be at cross purposes. Domestic radon is considered a major public health concern by the government and by many scientists. Yet the public is reluctant to take any action because of widespread apathy. The Yucca Mountain radioactive waste repository in Nevada, for example, continues to remain a contentious issue because experts and lay persons don't see eye to eye on safety, shared risk, and other issues. Experts and the DOE (the federal agency responsible for Yucca Mountain operations) argue that Yucca Mountain is a safe, geologically stable repository. Extensive testing and analyses of the Yucca Mountain site are continually referred to by the government to buttress its position. Transportation of waste to Yucca Mountain is also safe according to the federal government. Transportation casks are designed to withstand a wide variety of damage tests that far exceed what would be expected in a transportation accident. But lay groups see things differently. The public argues that there is no guarantee of safety, particularly since Yucca Mountain is a permanent repository and it is unknown whether toxic chemicals and radionuclides might leak into the environment over the next 10,000 years. Further the public believes that transportation accidents will occur because of the large number of expected shipments and long transportation distances. The public is concerned that if a transportation cask is compromised, disastrous consequences may result if large populations are exposed. Even if a cask remains intact as a result of an accident, there is concern that the public may be exposed to unacceptable radiation levels from radioactive waste within the cask. Since all shipments will be transported through Nevada, local residents think it is unfair that they should shoulder a disproportionate burden of risk.

It should be emphasized that reconciling expert and lay views is not a problem unique to risk management. Few scientific issues have been as contentious as human embryo research, especially research with human embryonic stem cells and cloning. While such research holds tremendous promise for therapies to manage diseases such as Alzheimer's disease, Parkinson's disease, and spinal cord injuries, it has met with opposition from individuals and groups who believe it is morally impermissible to destroy embryonic human life. At the center of the debate are competing views advocated by experts, lay groups, and religious groups — in this case about the meaning of life.

Contentious issues and competing views are not going to be resolved until parties are willing to factor differing perspectives into their own thinking. How parties choose to mediate conflicting views at the level of public policy will determine whether meaningful resolution can be achieved.

MANAGEMENT STRATEGIES

Radiological and chemical risks are managed differently. In radiological protection a hybrid dose-risk-based system is used.[33] The regulatory limit establishes a ceiling, and using the ALARA approach, doses are reduced to levels as low as possible given social and economic constraints. In contrast chemical hazards are managed using a risk-based system. Because the EPA regulates so many different chemicals, risk is viewed as the only way to manage multiple agents. Risk goals of 10^{-6} to 10^{-4} are established under CERCLA, and using a bottom-up approach, risks are appropriately adjusted to achieve health protection using the best available technology (BAT) or other suitable management approach. Because of differences in risk-management approaches, conflicts arise in settings where chemical and radiological hazards coexist.[34]

In many ways radiological protection is more advanced and mature than chemical risk management. Radiogenic risks are based on a substantial database of human exposures. Epidemiological studies involve a wide range of doses, including doses of environmental and occupational significance. Although radiological risk uncertainties remain problematic at very small doses, we know risks fairly well at higher doses. In the case of chemical carcinogens, few agents have been extensively studied; there is no direct human experience except for a small number of agents. Accordingly, risks are less certain than radiological risks because of uncertainties introduced by cross-species extrapolation. Limited human data are available for most chemicals labeled as probable or possible human carcinogens. The primary evidence for labeling these agents as carcinogenic comes from experiments in nonhuman systems.

As Low As Reasonably Achievable (ALARA)

ALARA seeks to balance costs of dose management with health benefits achieved by dose reduction. Effectiveness of an ALARA program can be measured by evaluating temporal trends in collective dose.[35] If socioeconomic constraints remain constant over time, collective dose (corrected for changes in the size of the population at risk) should remain constant or decrease in an effective ALARA program. The ALARA approach in radiological protection evolved from the linear no-threshold (LNT) theory that has for many years served as a basis for radiation protection practice and regulatory decision making. Because LNT is a monotonic dose-response function, dose reduction over any dose interval theoretically results in risk reduction (the amount of risk reduction being dependent on the dose savings). The theory predicts that any residual dose of radiation, no matter how small, might cause cancer. LNT has been interpreted by some to mean that there is no safe dose of radiation and, accordingly, the goal of an ALARA program should be zero dose. In a radiological environment, zero dose is impossible to achieve unless there is a decision not to use radiological sources, and even then irreducible natural background radiation is still present.

The key to an effective ALARA program is identifying what is "reasonable" in terms of costs and benefits. Unfortunately there is no clear decision rule that can be applied across all radiological environments. What may be reasonable and acceptable in one setting may not be in another because of differences in cost constraints and site-specific requirements. Accordingly, ALARA decisions are often made on a case-by-case basis. With the focus on balancing costs and benefits in dose reduction, health protection programs may not adequately address distributive justice questions in the application of ALARA. For example, under ALARA individuals bearing technological risks may not reap the benefits of the technology. Individuals living in close proximity to an electric generating plant may get their electric power from another source.

Historically, ALARA programs have worked well in radiological protection. The excellent safety record in nuclear and radiological technologies over the years is due primarily to a radiological protection system that has kept pace with the rapid technological advancements in electric power generation, engineering, and medicine.

Best Available Technology (BAT)

BAT is closely related to ALARA. It is a risk-management strategy that is driven by cost constraints with dose reduction as an indirect outcome. The objective is to employ the best technology to reduce dose. BAT assumes that candidate technologies are commercially, available appropriately field tested, and available at costs that meet application and site requirements. In theory there is nothing to prevent a company from developing and implementing its own risk-management technology. However, testing and operational costs may preclude such an approach instead of BAT. Clearly, the effectiveness of any BAT program is knowledge of available technology options. BAT has been applied to control of carcinogens in the workplace under regulations promulgated by the Occupational Safety and Health Administration (OSHA) and to control pollutants under the Clean Air Act administered by the EPA. Perhaps the most successful application of BAT has been the control of water pollution.[36]

The Precautionary Principle

The precautionary principle is not an unreasonable proposition on its face.[37] If a technology or product might cause harm, then alternatives should be considered to avoid harm. The problem is in how this principle is interpreted and applied. In the extreme, the principle can result in removal of technologies from society that have clear benefits but for which risks are small but highly uncertain. Chapter 9 discusses cell phones and how implementation of the precautionary principle led to questionable public policy.

The precautionary principle is a relatively new concept in risk management. But imagine if it had been around when electricity or the automobile were first introduced. These technologies would have never been sustained had the precautionary principle been invoked. Important scientific achievements such as Roentgen's discovery of x-rays in 1895 would have never reached its full potential under precaution. Uncertainties about radiation risks would have prevented development and implementation of medical diagnostic applications of x-rays. Implementation of the precautionary principle introduces an interesting paradox. By considering precautionary measures, decision makers hope to cope with public fears about technological risks. However, by implementing the precautionary principle, the opposite might occur by lowering public trust in health protection and amplifying risk perceptions.[38]

The precautionary principle is often summarized by the phrase "better safe than sorry." It requires foregoing, postponing, or otherwise limiting a product or activity until uncertainty about potential risks has been resolved in favor of safety. Over the past decade, the precautionary principle has been incorporated into a series of international environmental agreements, perhaps most prominently the 1992 United Nations Rio Declaration on Environment and Development, which states: "Where there are threats of serious and irreversible damage, lack of full scientific certainty shall not be used as a reason for postponing cost-effective measures to prevent environmental degradation."[39] Various formulations of the precautionary principle have also been adopted into other international agreements such as the United Nations Framework Convention on Climate Change, the Montreal Protocol on Substances that Deplete the Ozone Layer, and the Cartagena Protocol on Biosafety.

Europe has been at the forefront of adopting the precautionary principle, and the European Community (EC) formally committed to implementing environmental policy in conformity with the precautionary principle in the 1992 Maastricht amendments to the EC Treaty. Individual European nations, most notably France, Germany, and the Scandinavian nations, have selectively begun to implement the precautionary principle in their national regulatory programs, as have some non European nations, including Australia and Canada.[40]

As presently formulated, the precautionary principle is ill defined and vague. There is no standard definition of the precautionary principle, and the many versions that do exist are inconsistent in important respects.[41] For example, compare the language of the Rio Declaration with the version put forward as a consensus statement by many proponents of the precautionary principle known as the Wingspread Statement: "When an activity raises threats of harms to human health or the environment, precautionary measures should be taken even if some cause and effect relationships are not fully established scientifically."[42] The precautionary principle adopted by the Rio Declaration applies only to "serious and irreversible" risks, whereas the version provided by the Wingspread Statement is not limited to any subset of risks. The Rio Declaration requires any action taken under the precautionary principle to be cost-effective, while the Wingspread Statement makes no mention of economic considerations. The Rio Declaration is stated in the negative, in that uncertainty should not preclude preventive action, whereas the Wingspread Statement imposes an affirmative obligation to act notwithstanding uncertainty.

These and many other inconsistencies between the many different versions of the precautionary principle are compounded by the ambiguity in any specific formulation of the precautionary principle. No version of the precautionary principle is clear on when the precautionary principle applies, and just as importantly, when it does not apply. For example, is the principle triggered by the magnitude of a risk, the uncertainty associated with that risk, or some combination of both magnitude and uncertainty? How much of each is necessary to trigger the principle? If the principle applies only to "serious" or "irreversible" risks, how are such risks defined? If the principle is not limited to serious or irreversible risks, how can the principle be applied in a considered and feasible manner, given that every product presents some risks in some scenarios?

What evidence is necessary to establish the necessary magnitude of risk or uncertainty? Can the unsubstantiated fears of one or more persons trigger the principle? Or does the suspicion of a potential risk have to be supported by credible scientific evidence? What if there is some scientific evidence of a potential risk, but the total body of available evidence weighs against the existence of a significant risk? Who makes the decision on whether the evidence is sufficient to meet the standard for triggering the principle?

What types of "precautionary measures" should be taken when a sufficient threat exists? Should the precautionary measures be proportional to the magnitude of the "threat"? If the precautionary principle requires blocking development of a product until sufficient safety data on that product is available, what is required before the product is permitted to move forward? If the available evidence indicates the potential existence of some risk, what level of risk, if any, is acceptable to allow the product to proceed? What factors can be considered in determining whether the product should go forward? For example, can the economic benefits of the product be considered? Are the health and safety benefits of products considered?

These unanswered questions create substantial uncertainty about the applicability and requirements of the precautionary principle to any given risk. The European Union (EU) has made the most concerted attempt to try to reduce some of these uncertainties and provide some concrete guidance on the application and meaning of the precautionary principle. In particular, the 29-page communication on the precautionary principle issued by the European Commission (EC) in February 2000 provides the most detailed guidelines on the precautionary principle to date.[43]

The EC Communication provides some guidance on when recourse to the precautionary principle is triggered. The communication defines the precautionary principle as a risk-management tool, to be applied only after a scientific evaluation of the available risk data (i.e., risk assessment). The communication describes two outputs from this risk assessment that are necessary to justify recourse to the precautionary principle. The risk assessment must identify potentially negative effects resulting from the product or activity, and the available scientific data must be so insufficient, inconclusive, or imprecise to make it impossible to "determine with sufficient certainty the risk in question." A political decision is then required to determine whether any precautionary action is appropriate, based largely on the level of the risk that is acceptable to the communities in which the risk is imposed.

The precautionary principle is implicit in existing radiation safety practice but is not explicitly required. The ALARA philosophy is used to minimize the radiation dose in occupational and environmental settings with appropriate considerations for social and economic costs. When used appropriately, the ALARA philosophy balances the public health goal of maintaining doses as low as possible against economic and other costs of achieving specific dose targets.

Is a more stringent approach to radiation protection premised on the precautionary principle necessary and appropriate? Ionizing radiation does not meet the criteria identified by the EC Communication for recourse to the precautionary principle. In the first place, the existing scientific database for radiation is neither inadequate nor imprecise, requirements identified by the EC for triggering application of the precautionary principle. To the contrary, ionizing radiation is one of the most thoroughly studied human carcinogens. Health effects data at doses below 100 mSv are available from a number of published studies.[44] Low-dose risks are difficult to detect in epidemiological studies because of the large background rate of cancer and the fact that radiogenic cancers are clinically indistinguishable from cancers that arise from most other causes. Although low dose and low dose-rate epidemiological studies with very large populations may have sufficient statistical power to detect radiogenic risks, the use of low doses makes clear demonstration of radiation effects difficult. Other factors including the "healthy worker effect," contributions of possible confounding influences of chemicals and other toxic agents in the workplace, accuracy of dose assessment, mortality follow-up, and various lifestyle factors (e.g., smoking histories) may also cloud interpretation of data on radiogenic risk. Notwithstanding these uncertainties, the scientific database is sufficiently rich and robust to guide policy decisions without recourse to the precautionary principle.

Perhaps even more critical to the issue of whether the precautionary principle should apply to ionizing radiation is the question of acceptable risk. The precautionary principle should be triggered only by risks with the potential to impose unacceptable consequences. This inquiry necessitates establishing a level of risk that is acceptable (or perhaps trivial), below which neither regulatory intervention nor the precautionary principle is warranted.

Application of the precautionary principle is neither necessary nor appropriate for radiation protection given existing protections and policies in place. Even if the precautionary principle were applicable to ionizing radiation, many of the actions based explicitly or implicitly on the precautionary principle are inconsistent with the policies in the EC Communication governing application of the principle. For example, the principles of proportionality and cost-benefit evaluation argue against regulatory action for very low radiation exposures. This guidance appears inconsistent with some extreme and inappropriate applications of ALARA (premised on the precautionary principle) in which doses are reduced to the lowest levels possible (if not zero) with little, if any, benefit-cost considerations.

This overly precautionary approach to radiation protection leads to substantial economic expenditures for a minimal public health benefit and promotes public fear of radiation by fostering the idea that any dose of radiation is potentially harmful.

For example, the EPA set an annual individual-protection standard of 0.15 mSv in its final ruling on radiation standards for Yucca Mountain.[45] There is little evidence that doses in the range of natural background (approximately 1 mSv per year excluding contributions from radon gas and its progeny) are harmful to the public health or the environment. The EPA standard for Yucca Mountain is so low that it is within the *variation* of natural background radiation levels in the U.S.

In radiation safety, implementation of the precautionary principle is unnecessary. Sufficient information about public health risks of radiation exposure is known at radiation doses relevant in occupational and environmental settings. Moreover, any residual risks remaining after a prudent application of the existing ALARA policy would likely be in the acceptable risk range. One of the greatest ambiguities about the precautionary principle is the failure of its advocates to specify what level of risk is acceptable. Given that every product and human activity has the potential to create some risks, criteria such as acceptable risk thresholds are necessary to limit its applications to certain categories of risk. The alternatives are for the precautionary principle to apply to every risk, which is both impractical and imprudent, or for the principle to be applied in an arbitrary manner to some risks.

RISK–RISK TRADE-OFFS AND UNINTENDED CONSEQUENCES

Efforts to reduce a particular (target) risk often result in introducing or enhancing countervailing risks. Risk-management decision making must include consideration of the possibility that attempts to reduce one risk may increase the probability of others. Managing these unintended risks may be more challenging and serious than the efforts to control the target risk. When disproportionate attention is paid to one risk, the chances of unintended consequences occurring as a result of managing that risk may be elevated. The risk trade-off problem can be magnified when: (1) the target risk is very small because the importance of the countervailing risk as compared to the target risk is likely to increase, (2) there is no effective alternative to the banned or restricted activity (e.g., banning use of artificial sweeteners to reduce cancer risks might increase health risks in diabetics forced to use sugar), and (3) fear or excessive concern for a particular risk drives decision making.[46] Fear of x-rays may lead patients to avoid needed medical tests that could be of value in diagnosing disease. Doing nothing raises the risk of not identifying a disease in its early stages when the probability of therapeutic success may be highest. Proponents of the precautionary principle argue that we should avoid engaging in activities for which little risk information is available. By avoiding the technology and its attendant risks, other risks are created or enhanced because alternative technologies are utilized or no action is taken.

Risk–risk trade-offs are seen everywhere — from personal everyday decisions to political decisions on the world stage. One of the clearest examples of balancing target and countervailing risks is in medical therapy, particularly for cancer. High-dose radiation and chemotherapy can effectively manage many types of cancers (target risk) but in doing so they also can damage surrounding healthy normal tissues (countervailing risks). The dose prescription is constrained by the countervailing risks.

If the prescribed dose is too high, the tumor may be cured (the target risk has been successfully managed), but there may be unacceptable normal tissue injury (countervailing risks have been poorly managed). Alternatively, if the prescribed dose is too low, there may be no normal tissue injury but the patient will also not be cured of the disease. Oncologists are guided by two objectives in curative therapy: first treat the patient so that there is an acceptable probability of cure; second treat the patient so that there is an acceptable probability of normal tissue complication. Meeting these conflicting objectives implies that the physician prescribes cancer therapy agents within a narrow range of doses.

The introduction or exacerbation of countervailing risks in risk-management decisions involving very small target risks may have substantial economic, social, and environmental costs. These unintended consequences often require that original risk-management decisions be reconsidered. Trade-off risk typology is determined by the nature of the countervailing risk and the individuals or populations affected. The following examples illustrate the four main types of risk–risk trade-offs.[47]

Risk Offset

In a risk offset scenario, the target and countervailing risks are the same (e.g., cancer) and occur in the same population. Risk offsets are seen in efforts to reduce pesticide-related hazards to consumers and farm workers. Offset risks arise when substitute pesticides or pest-control practices partially or completely offset the reduction in the target risk.[48] Radiation and chemotherapy for cancer are another example. Ionizing radiation and some cancer drugs (e.g., cyclophosphamide) are effective cancer treatments but also can cause cancer. The risk-offset problem is seen primarily in pediatric cancer patients. Children successfully treated for cancer (target risk reduced) live long enough to express an increased risk of cancer decades later (countervailing risk).

Risk Substitution

Risk substitution is seen when different target and countervailing risks occur in the same population. The avoidance of air travel after 9/11 is a good example. Visions of aircraft crashing into the World Trade Center, the Pentagon, and in rural Pennsylvania heightened fears of air travel and drove many Americans to hit the highways because of fear of flying. Eliminating the risk of flying was substituted by the greater risks of traffic fatalities. The number of Americans who lost their lives on the road by avoiding the risk of flying was higher than the total number of passengers killed on the four fatal flights. Compared to the average number of fatalities for September, October, and November in the years 1996–2000, an additional 353 people died in car accidents in October, November, and December 2001. There were 266 fatalities on the four ill-fated aircraft on September 11, 2001.[49]

The controversy over farmed salmon in the diet is another example of risk substitution. Salmon is high in omega-3 fatty acids and eating it more than once a week may decrease one's risk of a heart attack by 30%. However, farmed salmon also contains polychlorinated biphenyls (PCBs) that may increase the risk of cancer at high concentrations. The health benefit (i.e., reduced risk of heart disease) from eating salmon must be weighed against any potential cancer risk from PCBs.[50]

Risk Transfer and Risk Transformation

A risk transfer occurs when the target and countervailing risks are the same but occur in different populations. A risk transformation occurs when different target and countervailing risks occur in different populations. Use of drugs contraindicated in pregnancy is an example. Prior to its withdrawal from the market in 1961, thalidomide was an effective sedative for the control of morning sickness in pregnant women. However, the drug also had teratogenic activity, causing specific developmental anomalies in the embryo. Another example involves banning of organophosphate pesticides. The ban eliminated neurotoxicity risks in consumers from chronic exposure to the pesticide in the diet. But the ban also introduced countervailing risks of adverse health effects on farm families due to crop failures and decreased farm income.[51]

Evaluating risk–risk trade-offs complicates the risk-management decision process. Without appropriate consideration of countervailing risks and their impacts, a risk-management decision may be less beneficial (or more detrimental) than originally anticipated and entail significant costs. The decision to avoid aggressive cleanup of Par Pond at the DOE Savannah River Site is an example of how considering risk–risk trade-offs can avoid costly remediation and unintended consequences and at the same time provide effective solutions to environmental and public health problems. Owing to concerns about dam integrity, the DOE drained water from the reservoir to reduce risk of dam failure. This action resulted in exposure of human populations to radioactive cesium contaminated sediments in the exposed reservoir bed. Serious consideration was given to excavating and trucking the radioactive sediments off-site at a cost of more than $4 billion to eliminate the theoretical public health risk. After analysis of alternative strategies, the DOE decided to repair the dam at a cost of about $12 million and refilled the reservoir without disturbing the sediment at the bottom of the reservoir. This action entailed significant cost savings, protected the public health, and preserved the environment.[52]

Regulatory analysis for many agents including pesticides does not readily allow for evaluation of risk–risk trade-offs. Because of uncertainties in exposure, risks, and benefits, alternative strategies may be difficult to evaluate because risk trade-offs cannot be easily quantified. In the Par Pond example above, decision making was facilitated by substantial cost differences in risk-management strategies. In medicine, risks and benefits are usually well known before drugs are approved by the U.S. Food and Drug Administration (FDA).[53]

CHALLENGES

Risk management is a challenging yet frustrating exercise because people look at the process from differing perspectives. A satisfactory risk-management solution to a particular environmental or public health problem may be difficult to achieve because impacted and interested stakeholders have different concerns, different agendas, and different goals.

The lack of clear rules to guide decisions on when and how to apply risk-management approaches only makes the problem worse. Although ALARA has been successfully used in radiological protection for decades, decision rules on what constitutes acceptable dose are lacking. Until such rules are put into place, radiological risk-management practices will lack uniformity. If the precautionary principle is to serve any useful purpose in risk management, decision makers cannot ignore the need to weigh benefit against harm, or to consider the effect of taking no action. In the European Union, the precautionary principle has been used in an arbitrary and capricious fashion because there has been no consistent or rational basis for applying it. Again clear rules governing how and when to apply the principle are lacking.[54]

The demarcation of an acceptable risk range to which risk management no longer applies would almost certainly exempt many low-level exposures to radiological and chemical agents from further concern. Defining "acceptable risk," which is critical for evaluating the need for recourse to risk-management strategies including ALARA and the precautionary principle, would effectively address two major interrelated problems: economic costs associated with reduction of trivial risks and the idea that very small doses of carcinogens are harmful. Specifically, for radiological protection:

- Professional and scientific organizations (for example, the Health Physics Society, American Association of Physicists in Medicine, and the American College of Radiology) should issue official positions (preferably jointly) defining acceptable dose of radiation for unintentional radiation exposures or intentional exposures such as medical or dental x-rays for which there is a clear benefit.[55] If clinicians, engineers, and scientists are unwilling to declare what they believe to be an acceptable dose, the public will continue to be justified in its belief that no dose is safe. Scientists and technical professionals have an obligation, either individually or as a community, to provide reasonable and responsible guidance to the public because of social responsibility and because most research and development activities are supported by public funds. Guidance should be viewed as just that — guidance. Experts' views should be considered as one of many perspectives that the public may use in evaluating and managing risks.
- The U.S. government should revisit its abandoned below regulatory concern (BRC) policy. The federal government attempted to define an acceptable level of risk through a 1985 Congressional mandate to the U.S. Nuclear Regulatory Commission (USNRC) to establish a BRC policy.[56] The policy was to establish a framework whereby the USNRC would formulate rules or make licensing decisions to exempt from regulatory control those practices that have such low estimated health risks that further reduction of those risks would be unwarranted. In July 1990, the USNRC established a BRC policy.[57] Special interest groups opposed the policy on the grounds that it would lead to uncontrolled release of radioactive material. This view, coupled with the lack of consensus within the government regarding BRC risk levels, ultimately led to Congressional

revocation of the BRC policy in 1992.[58] Although there was some disagreement about BRC risk levels, federal agencies and the technical community generally agreed that a BRC-type policy was a worthwhile concept for effectively allocating and managing regulatory resources.

NOTES AND REFERENCES

1. Cigarette smoking is perhaps the best example. There is clear evidence from numerous epidemiological studies that reducing cigarette consumption leads to a reduction in cancer mortality (and also cardiovascular disease mortality). The New York City Department of Health has reported an 11% reduction in cigarette consumption from 2002 to 2003 and predicts that, as a consequence, about 30,000 premature deaths due to cardiovascular disease and some cancers will be averted. New York City Department of Health and Mental Hygiene, Office of Communications, *New York City's Smoking Rate Declines Rapidly from 2002 to 2003, the Most Significant One-Year Drop Ever Recorded*, May 12, 2004.
2. National Council on Radiation Protection and Measurements (NCRP), *Exposure of the Population in the United States and Canada from Natural Background Radiation*. NCRP Report No. 94. NCRP, Bethesda, MD, 1987.
3. Coglianese, C. and Marchant, G.E., Shifting sands: the limits of science in setting risk standards, *University of Pennsylvania Law Review,* 152(4), 1255, April 2004.
4. Ibid.
5. *Supra* note 3.
6. International Commission on Radiological Protection (ICRP), 1990 Recommendations of the International Commission on Radiological Protection, ICRP Publication 60, *Annals of the ICRP* 21(1–3), 1990.
7. Dose reduction in diagnostic radiology must not compromise the diagnostic value of the study. Key strategies to reduce dose include optimizing fluoroscopic time, reducing the number of films, eliminating the need for repeat studies and careful selection of x-ray exposure factors to maximize image quality. Operator training and performance are also important determinants of image quality. The U.S. Congress enacted the Mammography Quality Standards Act (MQSA) in 1992 to ensure that all women have access to quality mammography for the detection of breast cancer in its earliest, most treatable stages by requiring that mammography centers meet certain standards for machine performance, film processing, and staff training and performance. The legislation is concerned with any aspect of mammography, including the production, processing, and interpretation of mammograms and related quality assurance activities.
8. Fiorino, D.J., Technical and democratic values in risk analysis, *Risk Analysis,* 9, 293, 1989.
9. In providing guidance to the public regarding radon in homes, the EPA offers recommendations based on the level of radon concentration in air. EPA policy assumes that radon concentration is directly related to risk of lung cancer. Radon concentration in air is measured in Bqm^{-3}. If concentrations are below 75 Bqm^{-3} no action is recommended to homeowners to reduce concentration levels. Between 75 and 150 Bqm^{-3} the EPA recommends that homeowners consider fixing their home to reduce domestic radon levels. For homes with radon concentrations above 150 Bqm^{-3} EPA recommends that homeowners fix their homes. See EPA, *A Citizen's Guide to Radon* (2nd ed.), U.S. EPA Air and Radiation (ANR-464), U.S. Government Printing Office, Washington, DC, 1992.

10. Environmental Protection Agency regulations establish carcinogen dose limits to meet a target risk of 1 in 10,000. This target risk derives from CERCLA law. With the enactment of CERCLA in 1980, the Congress created the Superfund program authorizing the EPA to, among other things, clean up contamination at hazardous waste sites. In the case of cancer, the EPA considers the risk serious enough to warrant cleanup if the risk assessment indicates more than a 1 in 10,000 probability that exposure to the site's contaminants may cause an individual to develop cancer. The Comprehensive Environmental Response, Compensation and Liability Act (CERCLA) enables the EPA to protect the public health and environment from "releases" of hazardous (toxic) substances into the environment.
11. Human diseases associated with enhanced sensitivity to radiation include Ataxia telangiectasia (AT), characterized by enhanced x-ray sensitivity and Xeroderma pigmentosum (XP), characterized by increased sensitivity to solar ultraviolet radiation. AT is a rare, recessive genetic disorder of childhood that occurs in about 1 in 100,000 persons worldwide. The disease is characterized by neurologic complications, recurrent serious sinus and respiratory infections, and dilated blood vessels in the eyes and on the surface of the skin. Patients usually have immune system abnormalities and are very sensitive to ionizing radiation exposure. AT patients are at high risk of developing and dying of cancer, particularly leukemias and lymphomas. An estimated 1% of the U.S. population may be carriers for AT. See Sankaranarayanan, K. and Chakraborty, R., Impact of cancer predisposition and radiosensitivity on the population risk of radiation-induced cancer, *Radiation Research,* 156, 648, 2001 for an overview of current evidence for cancer predisposition and for increased radiosensitivity to certain cancers.
12. National Research Council, *Health Risks from Exposure to Low Levels of Ionizing Radiation,* BEIR VII Report, National Academies Press, Washington, DC, 2005.
13. Adverse effects require direct exposure of the developing embryo and fetus. Exposure of the fetus can occur following certain medical radiodiagnostic procedures, particularly involving the lower abdomen or pelvis. Chemical agents must cross the placenta in order to exert an effect.
14. Mossman, K.L., Radiation protection of radiosensitive populations, *Health Physics,* 72, 519, 1997; International Commission on Radiological Protection (ICRP), *Genetic Susceptibility to Cancer,* ICRP Publication 79, Elsevier Science, Oxford, 1998.
15. For ionizing radiation assigned share, AS, is defined as

$$\text{AS} = \frac{\text{risk due to radiation exposure}}{\text{baseline risk} + \text{risk due to radiation exposure}} \times 100\%$$

    For calculation purposes excess relative risk (ERR) = risk due to radiation exposure/baseline risk,

$$\text{AS} = 100\% \times \text{ERR}/(1 + \text{ERR})$$

    See U.S. Department of Health and Human Services, Report of the NCI-CDC Working Group to Revise the 1985 NIH Radioepidemiological Tables, 2003.
16. The AS was calculated using the National Institute of Occupational Safety and Health Interactive RadioEpidemiological Program (NIOSH-IREP v. 5.2.1) http://www.cdc. gov/niosh/ocas/ocasirep.html (accessed March 2006). AS depends on numerous factors including age at exposure, age at diagnosis, smoking history and radon exposure. The AS increases for never smokers and for domestic environments with very high radon concentrations.

17. Energy Employees Occupational Illness Compensation Program Act of 2000, as Amended, 42 U.S.C. § 7384 *Et Seq;*. Department of Health and Human Services, *Guidelines for Determining the Probability of Causation and Methods for Radiation Dose Reconstruction Under the Employees Occupational Illness Compensation Program Act of 2000,* Final Rule, 42 CFR Parts 81 and 82. The Act requires that decisions be based on the upper bound of the 99% confidence interval of the calculated AS value. The type of cancer, past health-related activities, the risk of developing a radiation-related cancer from workplace exposure, and other relevant factors must also be considered in the decision to compensate an eligible worker.
18. U.S. Department of Health and Human Services, Report of the NCI-CDC Working Group to Revise the 1985 NIH Radioepidemiological Tables, 2003; Johnson, RH., The radioepidemiological tables in radiation litigation, in Mossman, K.L. and Mills, W.A. (Eds.), *The Biological Basis of Radiation Protection Practice,* Williams & Wilkins, Baltimore, MD, 158, 1992.
19. Slovic, P., Perception of risk, *Science,* 236, 280, 1987.
20. Lowrance, W.W., *Of Acceptable Risk: Science and the Determination of Safety,* William Kaufmann, Inc., Los Altos, CA, 1976.
21. Salazar-Lindo, E. et al., The Peruvian cholera epidemic and the role of chlorination in its control and prevention, in *Safety of Water Disinfection: Balancing Chemical and Microbial Risks,* Craun, G.F. (Ed.), 403, ILSI Press, Washington, DC, 1993.
22. Bate, R., How precaution kills. The demise of DDT and the resurgence of malaria, in *Politicizing Science,* Gough, M. (Ed.), Hoover Institution Press, Stanford, CA, 261, 2003.
23. Bradley, M.O., For fearful residents, there was no script. *The Patriot-News,* March 31, 2004. http://www.pennlive.com/news/patriotnews/index.ssf?/news/tmi/stories/forfearful.html (accessed March 2006).
24. Cothern, C.R., Widespread apathy and the public's reaction to information concerning the health effects of indoor air radon concentrations, *Cell Biology and Toxicology,* 6, 315, 1990.
25. Gregg Easterbrook interestingly argues that the focus on smaller and smaller risks is a reflection of the decline of big risks. There is little support for this position in view of compelling evidence that the incidence of serious public health problems, including diabetes and obesity, has risen significantly in recent decades. See Easterbrook, G., *The Progress Paradox: How Life Gets Better While People Feel Worse,* Random House, Inc., New York, 2003.
26. Brodeur, P., Annals of radiation — The hazards of electromagnetic fields (Part I-Power lines), *The New Yorker,* 51, June 12, 1989; Brodeur, P., Annals of radiation — The hazards of electromagnetic fields (Part II-Something is happening), *The New Yorker,* 47, June 19, 1989; Brodeur, P., Annals of radiation — The hazards of electromagnetic fields (Part III-Video display terminals), *The New Yorker,* 39, June 26, 1989.
27. Oak Ridge Associated Universities, *Health Effects of Low-Frequency Electric and Magnetic Fields,* ORAU 92/F8, Oak Ridge Associated Universities, Washington, DC, 1992.
28. Fiorino, D.J., Technical and democratic values in risk analysis, *Risk Analysis,* 9, 293, 1989; Ledwidge, L., Moore, L., and Crawford, L., Stakeholder perspectives on radiation protection, *Health Physics,* 87, 293, 2004.
29. Proctor, R.N., *The Nazi War on Cancer,* Princeton University Press, Princeton, NJ, 1999.

30. Kraus, N., Maimfors, T., and Slovic, P., Intuitive toxicology: Expert and lay judgments of chemical risks, *Risk Analysis*, 12, 215, 1992; Barke, R.P. and Jenkins-Smith, H.C., Politics and scientific expertise: Scientists, risk perception, and nuclear waste policy, *Risk Analysis,* 13(4), 425, 1993.
31. Stelfox, H.T. et al., Conflict of interest in the debate over calcium-channel antagonists, *New England Journal of Medicine,* 338(2), 101, January 8, 1998; Levine, J. et al., Authors' financial relationships with the food and beverage industry and their published positions on the fat substitute olestra, *American Jornal of Public Health,* 93(4), 664, 2003.
32. Beckelman, J.E., Li, Y., and Gross, C.P., Scope and impact of financial conflicts of interest in biomedical research: a systematic review, Journal of the American Medical Association, 289, 454, January 22–29, 2003; Lexchin, J. et al., Pharmaceutical industry sponsorship and research outcome and quality: systematic review, *British Medical Journal*, 326, 1167, May 31, 2003.
33. Regulatory limits are expressed by a quantity referred to as effective dose, the product of the measured absorbed dose, a dimensionless radiation weighting factor and a dimensionless tissue weighting factor. The tissue weighting factor is actually a risk proportion. See International Commission on Radiological Protection (ICRP), 1990 Recommendations of the International Commission on Radiological Protection, ICRP Publication 60, *Annals of the ICRP,* 21, 1–3, 1990; Mossman, K.L., Restructuring nuclear regulations, *Environmental Health Perspectives,* 111, 13, 2003.
34. The EPA and the U.S. NRC have duplicative oversight of nuclear energy facilities. The agencies have failed to resolve issues associated with clean-up of licensed sites because of agency differences in legislative and regulatory risk management requirements.
35. Collective dose is the sum of all doses received by all members of a population at risk. For large populations it is the product of the average population dose and the number of individuals in the population. See National Council on Radiation Protection and Measurements, *Principles and Applications of Collective Dose in Radiation Protection,* NCRP Report No. 121, NCRP, Bethesda, MD, 1995.
36. Wilson, R. and Crouch, E.A.C., *Risk Benefit Analysis*, Harvard University Press, Cambridge, 2001.
37. The discussion of the precautionary principle is based on Mossman, K.L. and Marchant, G.E., The precautionary principle and radiation protection, *Risk: Health Safety & Environment,* 13, 137, 2002.
38. Wiedemann, P.M. and Shutz, H., The precautionary principle and risk perception: Experimental studies in the EMF area, *Environmental Health Perspectives,* 113, 402, 2005.
39. United Nations Conference on Environment and Development, Rio Declaration on Environment and Development, United Nations, New York, 1992.
40. France's National Assembly has taken steps to incorporate the precautionary principle into the country's constitution. See precaution versus principles, *Nature,* 429, 6992, 585, June 10, 2004.
41. Sandin, P., Dimensions of the precautionary principle, *Human and Ecological Risk Assessment,* 5, 889, 1999.
42. Raffensperger, C. and Tickner, J.A., *Protecting Public Health and the Environment: Implementing the Precautionary Principle,* Island Press, Washington, DC, 1999.
43. European Commission, *Communication for the Commission on the Precautionary Principle;* 2000.
44. There are several large cooperative epidemiological studies ongoing. Many of these studies involve nuclear workers with study populations approaching 100,000 subjects.

Average doses to subjects are less than 10 mSv. The single most important study, however, is the Life Span Study of the Japanese survivors of the atomic bombings in 1945. See United Nations Scientific Committee on the Effects of Atomic Radiation, *Sources and Effects of Ionizing Radiation,* UNSCEAR 2000 Report to the General Assembly, with scientific annexes, United Nations, New York, 2000.

45. Environmental Protection Agency, Public health and environmental radiation protection standards for Yucca Mountain, NV, final rule. *Federal Register,* 66, (114), 32073, 2001.
46. Graham, J.D. and Weiner, J.B., Confronting risk tradeoffs. In Graham, J.D. and Weiner, J.B. (Eds.), *Risk vs. Risk,* Harvard University Press, Cambridge, 1, 1995.
47. Ibid.
48. See Gray, G.M. and Hammitt, J.K., Risk/risk trade-offs in pesticide regulation: an exploratory analysis of the public health effects of a ban on organophosphate and carbamate pesticides, *Risk Analysis,* 20, 665, 2000.
49. Gigerenzer, G., Dread risk, September 11, and fatal traffic accidents, *Psychological Science,* 15, 286, 2004.
50. Hites, R.A. et al,. Global assessment of organic contaminants in farmed salmon, *Science,* 303, 226, January 9, 2004.
51. Gray, G.M. and Hammitt, J.K., Risk/risk trade-offs in pesticide regulation: an exploratory analysis of the public health effects of a ban on organophosphate and carbamate pesticides, *Risk Analysis,* 20, 665, 2000.
52. Whicker, F.W. et al., Avoiding destructive remediation at DOE sites, *Science,* 303, 1615, March 12, 2004.
53. The FDA requires extensive drug testing to characterize benefits and risks before approval of the drug for sale. A drug may be withdrawn from the market by the manufacturer or the FDA if widespread use identifies previously undetected side effects.
54. Marchant, G.E. and Mossman, K.L., *Arbitrary and Capricious: The Precautionary Principle in the European Union,* AEI Press, Washington, DC, 2004.
55. The Health Physics Society has revised its Position Statement *Ionizing Radiation-Safety Standards for the General Public* (revised June 2003) to include the recommendation to support "the establishment of an acceptable dose of radiation of 1 mSv/y (100 mrem/y) above the annual natural radiation background. At this dose, risks of radiation-induced health effects are either nonexistent or too small to be observed." http://hps.org/documents/publicdose03.pdf (accessed March 2006).
56. Low Level Radioactive Waste Policy Amendments Act of 1985. 99 Stat.1842; Public Law 99-240.
57. U.S. Nuclear Regulatory Commission, Below regulatory concern, Policy statement, Federal Register, 55, 27522, 1990.
58. Energy Policy Act of 1992, Section 2901, 1992.

6 Misplaced Priorities

The societal decision to manage a particular risk is a complex process that is anchored in prioritization. By prioritization we mean the ordering of risks in terms of their size (i.e., probability of occurrence and health or environmental consequences), public perception, controllability, and socioeconomic and political impact. Prioritization reflects society's valuation of risks. Prioritization does not correlate completely with risk magnitude or severity. Resources are limited to manage risks, and society's decisions regarding priority provides some rationale for how resources should be allocated.

The goal of risk management is to reduce or eliminate risks. Once a risk assessment has been completed, decisions need to be made by regulators and policy makers about whether the risk should be managed, and if so what management strategies should be implemented. Not every risk requires or demands reduction to protect public health and the environment. Which risks are deemed important enough to be managed is a complex societal process that transcends science.

It would be reasonable to assume that prioritization is based primarily on scientific information that quantifies consequences and probabilities. It makes sense to think that the largest risks should be the ones that are managed first. Products and activities that cause the greatest harm ought to be at the top of the priority list. But in reality that is not the case. Science is influential only at the front end of the prioritization process when risk assessment is performed. After that, the influence of scientists and science diminishes quickly. Public policy, court opinions, and arguments and perspectives from stakeholders drive the risk-management agenda from there. Scientific assessments are important, but they are minor drivers in the prioritization process. The result is that some minor risks (in terms of the impact of the product or activity on public health and the environment) garner the most attention and resources while other more serious risks are left relatively unmanaged.

This chapter explores several key questions in the prioritization problem. Why do inequities in risk management exist? Why do we allocate substantial resources to manage risks (e.g., pesticide levels in food) that pose little or no public health hazard? Yet well-known risky activities such as cigarette smoking are devoid of comprehensive regulation. What and who determines which risks are important? What is the impact of allocating resources to manage risks that have relatively little impact on the public health? What is the cost of misplaced priorities?

An important theme of this book is the need to balance risk with benefits and with competing risks when deciding which risks should be managed and which risks should be ignored. Risk analysis is more than just evaluating an individual risk. Technological risks are not isolated and need to be considered in light of benefits of the product or activity and possible countervailing risks. Society should focus on management of risks for which there may be substantial gain in public health benefit.

The idea that we need to rethink how we prioritize risks is not new. What is new in this book is the notion that analyzing and discussing individual risks without regard to the presence of other risks is inappropriate. Often risks that may appear to be important when considered in isolation become less significant when compared to other risks in the environmental or occupational setting.

PRIORITIES AND REALITIES

Environmental regulations to protect the public health are primarily concerned with control of cancer. Cancer mortality is the principal health endpoint of concern to agencies such as the U.S. Environmental Protection Agency (EPA) and the U.S. Nuclear Regulatory Commission (U.S. NRC). But there is a clear distinction between what is controlled and the magnitude of the health risk. Cigarette smoking and certain dietary factors contribute significantly to the cancer mortality burden, but these factors are not regulated in any comprehensive way. Pesticides and pollutants contribute much less to the cancer burden but are strictly regulated.

A number of cancer risk factors have been well characterized. Twenty-five years ago Doll and Peto, two well-respected British epidemiologists, evaluated the epidemiology literature to determine the contributions of known risk factors to the U.S. cancer burden (Table 6.1).[1] The original study was published in 1981 before the HIV/AIDS pandemic. Although HIV contributes to the cancer burden, the relative ranking of the risk factors shown in Table 6.1 has not changed since 1981. Diet and cigarette smoking still account for about two-thirds of cancers. If infectious agents

TABLE 6.1
Ranking Cancer Risks

Risk Factor	Percent of All Cancer Deaths	Annual Number of Deaths Attributable to Factor	Priority
Diet	35	210,000	Low
Tobacco	30	180,000	Medium
Infections, reproductive and sexual behaviors	15	90,000	Low
Occupational exposures (chemicals, radiation)	4	24,000	High
Alcohol	3	18,000	Medium
Geophysical factors (including natural background)	3	18,000	Low
Pollution	2	12,000	High

Source: Percent of all cancer deaths from Doll, R. and Peto, R. The causes of cancer: Quantitative estimates of avoidable risks of cancer in the United States today, *Journal of the National Cancer Institute,* 66, 1191, 1981.

and sexual behaviors are included, about 85% of cancer with known risk factors is accounted for.

The risk factors in Table 6.1 are listed in order of their contribution to the U.S. cancer burden. The percent of cancer deaths attributable to each risk factor is a central estimate and does not reflect the level of uncertainty. For some estimates the bounds of uncertainty may be an order of magnitude or more. Numerous epidemiological studies were evaluated to derive these estimates. It is not surprising that variability among studies is substantial. The annual number of cancer deaths shown in the third column assumes a total cancer burden of 600,000 deaths annually. The number of deaths attributed to a particular risk factor is probably overestimated because the total cancer burden includes deaths for which no known risk factors have been identified.

The ranking of the Doll-Peto risk factors does not fully explain the entire cancer burden. For example, prostate cancer has a very high incidence rate in the U.S., striking about 180,000 males annually. But very little is known about what causes the disease. No specific causal factors have been identified. Some cancers like female breast cancers have a defined genetic component, but environmental factors are not fully understood. Ionizing radiation is a known risk factor but only at high doses; doses typically encountered in mammography have not been associated with elevated cancer risk. It is generally believed that heritable factors account for about 10% of cancers in Western populations. This, however, does not mean that the remaining 90% of cancers are caused by environmental factors. The nongenetic component of cancers includes a wide spectrum of interacting elements that may be expected to vary over time with changing social and economic conditions. These include population structure and lifestyle factors such as diet, reproductive behaviors, and certain types of infection (e.g., human papillomavirus).

Cancer risk factors may be prioritized as low, medium, or high depending on the degree of regulatory control. This rather subjective classification reflects the degree of regulatory oversight. Cigarette and alcohol consumption are considered medium priority. There is limited regulatory control over smoking and alcohol. Cigarette and alcohol consumption are regulated for minors but not adults, although in some U.S. municipalities (e.g., New York City, San Diego, CA, and Tempe, AZ) smoking bans in restaurants and other public buildings have been legislated. Many jurisdictions have established limits on business hours for serving alcohol. Conspicuously posted signs in bars warn pregnant women that alcohol and cigarette smoking are hazardous to the unborn child's health. Yet there is no restriction on consumption. Alcohol consumption is regulated indirectly in a broad spectrum of local, state, and federal transportation and workplace laws. Taxes and increased life insurance premiums are also indirect regulatory controls by differentially penalizing individuals. Dietary factors have low priority and are not regulated in any specific ways except for the requirement that nutritional and other dietary information be provided to consumers. Exposure to viruses through reproductive and sexual behaviors is also afforded low priority because of the government's limited interest in interfering with personal lives. The government supports education campaigns and disease prevention programs, but there are no regulations to control risky behaviors per se. Exposure to chemicals, radiation, and pollutants in occupational and environmental settings is strictly regulated through establishment of dose and risk limits and approved risk-assessment and -management strategies and has therefore been labeled high priority. Although its collective contribution to the cancer burden is a tiny fraction of the total burden, substantial costs are incurred by government, industry, and the public to manage these risks. A small number of cancers is attributable to naturally occurring factors that vary by geography and geology. In the U.S., for instance, individuals living in the Rocky Mountain areas of Colorado and New Mexico are subject to natural background radiation levels that are about twice as high as the levels in the middle Atlantic region. These differences are due to the radionuclide composition of Earth's crust and altitude. Control of geophysical factors is low priority; there is nothing that can be done to alter natural background radiation levels except to relocate to lower background areas.

FACTORS IN PRIORITIZATION

Prioritizing risk is a complex social process that involves consideration of scientific assessments of probability and severity of the event or agent exposure; social factors including individual and population perception of risk; capacity to manage the risk; political factors including the influence of special-interest groups and other stakeholders; and legal actions. Table 6.2 summarizes these factors and their likely impact on prioritization. Some factors tend to increase risk priority while others are either neutral or decrease priority. In reality, the result of the complex, dynamic interplay of factors is that prioritization is not congruent with measured probability and severity. Activities and agents that contribute significantly to the degradation of public health and the environment are not given the attention needed to control them better.

The tension that exists between experts and the public concerning technological risks stems in part from different views about risks and how they should be managed. Scientists and other experts tend to be utilitarian and positivist in their views. As utilitarian, scientists and technocrats believe the most serious risks should be given highest priority because control will result in the greatest good for the greatest

TABLE 6.2
Influences on Prioritization

Prioritization Factor	Prioritization Impact	Comment
Scientific evidence of risk	Increase priority	Use of conservative assumptions in risk assessment likely to overestimate true risk; may be viewed with skepticism by nonscientists
Public perception	Increase priority	Overrespond to small technological risks
Risk management capacity	Decrease priority	Underrespond to risks associated with personal behaviors; underrespond to natural risks because they may not be controllable
Court actions	Neutral	No trend toward overresponse or underresponse to risks
Stakeholder groups	Increase priority	Tend to overrespond to risk

number of people. The public acts as a community of individuals whereby risks and their control are contextualized at the personal level. In this view even very small risks may be considered serious if personal safety is thought to be threatened. The public harbors negative feelings about technologies and their risks. It is difficult for the public to place risks into perspective, particularly when benefits of technology are not clear. The public gets its information about risks from a variety of sources. Some of these sources are more trustworthy than others. Information that suggests that risks may be high are considered more acceptable to the public than evidence suggesting risks are small or zero. The public puts more trust in information agents (e.g., scientists) who are not viewed as biased. Industry scientists are considered less credible than scientists who serve on impartial government panels.

Scientific Evidence

As discussed in earlier chapters the process of evaluating risks is highly uncertain for many agents and events. For most carcinogens we do not know what the risk is to humans exposed to concentrations or doses of agents that may be encountered in occupational settings or in everyday life. Scientists and risk assessors usually adopt a conservative posture in estimating risk that is inherently precautionary.

Scientific assessment is a rational process that allows for ranking of risks according to their severity and probability of occurrence. Although scientific evidence is not the major driver in the risk-prioritization process, it is nevertheless the key initiating step. The listing of saccharin as a human carcinogen was based on limited scientific evidence in animals and no human evidence. It was subsequently delisted because of reassessment of the same data.

The listing and later delisting of saccharin as a carcinogen is an example of how science and politics are inexorably linked. The public funding of science through federal agencies such as the National Institutes of Health (NIH), National Science Foundation (NSF), and the National Aeronautics and Space Administration (NASA) is driven by political agendas. Through Congressional appropriations and executive branch directives and policies, funding priorities are established for virtually every scientific endeavor, including biomedicine and biotechnology, space exploration, and energy research. Politicians can intervene in the science enterprise by diverting support from one area to other areas. This can have a chilling effect by truncating progress in areas of investigation with real promise. Public officials use science to advance their political agendas.

Risk assessment can be tricky when the hazard is very small. Little direct scientific evidence may be available to support risk assessment at doses encountered in environmental or occupational settings. Whatever evidence is available is usually suggestive because risk data are derived (through analytical processes including dose and species extrapolation) rather than observed directly. Since risks cannot be tied directly to exposures, scientific evidence becomes contestable by interested and affected parties.

The public view of scientific risk assessment is that it is a cold, objective process that ignores important subjective, qualitative characteristics of risk that are important to the lay public. Scientific analysis often discounts contextualizing events and agent exposures in terms of benefits and social distribution of risks and benefits. Scientific assessments usually do not consider important social dimensions of risk such as equitable distribution of risks and benefits. Although a risk may be minor in a scientific context, it may nevertheless be unacceptable to some groups or communities.

Science also provides no basis for comparing disproportionate types of risks. Health outcomes may vary significantly for different agent exposures. Cancer mortality is a frequent health endpoint in risk assessment, but cancer deaths are not all the same. Risk analysts are now using methods employing years of life lost to measure quality of health impacts. Such methods recognize the difference between a child and an 80-year-old person dying of cancer. Both deaths are regrettable, but in one sense the child's death is more regrettable because of the number of years lost. Risk analysis does not fully account for differences in cancer types as contributors to mortality. The probability of dying from lung cancer is about nine times higher than from thyroid cancer; the diseases are characterized by entirely different clinical courses and treatment strategies.

Scientific evidence has important limitations that impact its utility in risk assessment. At small doses, measurement of no risk does not necessarily imply the absence of risk. Measurements of risk at very small doses are inherently uncertain because of methodological limitations and the nature of the disease being studied.[2] Science cannot answer the question of whether a particular exposure is "safe." By safe we mean what level of risk is deemed acceptable. Acceptability of a given risk transcends strict scientific analysis and involves judgments in the social, political, and economic arena.

Public Perception of Risks

Hard numbers about risks do not necessarily guide the public toward rational decisions. Nonscientists tend to overestimate risks that are unfamiliar, that may threaten future generations, and that may generate vivid images of past horrific events (e.g., the Three Mile Island and Chernobyl nuclear power plant accidents).[3]

The public's views of risk do not necessarily coincide with reality. When groups of individuals were asked to rank risks associated with various activities and technologies, significant differences in rankings were observed. The concept of risks means different things to different people. Expert rankings correlated well with technical estimations of fatalities. Lay people including college students and League of Women Voters ranked risks differently and depended more on other risk characteristics (e.g., catastrophic potential, long-term consequences) as a basis for risk judgments.[4]

The public has its own view of what risk is. Collection of scientific data and facts on potential hazards does little to calm the fears of the public. Concerns about potential radiation health effects of cell phones have been raging since the late 1980s. Numerous studies analyzing the scientific data indicate that risks are essentially nonexistent, but these authoritative studies have not put an end to the public controversy. The concern is public attitudes and reactions to perceived, rather than actual, risks.[5]

What does risk perception have to do with how we prioritize risks? Which risks are importants and which ones are insignificant? The public fears many risks that are inconsequential, and limited public funds are expended toward managing risks that contribute very little to public health problems. In spite of the fact that life expectancy in America has increased considerably in the last 100 years, the public has become increasingly concerned about health consequences from minor risks. The public is almost obsessed about small risks, and technology-associated risks are particularly scrutinized. Risks from natural products (e.g., naturally occurring carcinogens in foods) or natural events (e.g., tornadoes and hurricanes) are not viewed with the same disdain as technological risks.

The public fixates on very small risks in part because of television images and other mass media efforts that tend to sensationalize and exaggerate risk. Politicians are also responsive to sensationalism and public reaction. Effective journalism is based on the principle that bad news is good news but good news is no news. To sell papers, the public is constantly bombarded with reports about small probability events and their terrifying consequences. As a consequence the public perceives many small risks as serious threats and demands regulations to control these risks that are by their very nature difficult to reduce. Engineering controls are already adequate to manage these risks, but the public demands further reductions that are impossible to achieve without extraordinary expenditures of resources that could be better put to use controlling other more significant public health and environmental risks.[6]

There is a substantial literature on factors that impact risk perception.[7] One set of factors is concerned with cognition and individual experiences. The second is concerned with characteristics of the risk itself. How the risk is framed or presented is particularly important in shaping perception. It makes a significant difference whether risks are expressed as percentages or as frequencies (e.g., 25% versus 1 in 4) and whether outcomes are expressed as gains or losses (e.g., 75 out of 100 people died versus 25 out of 100 lived).[8] Individuals tend to be risk averse when the problem is expressed as a choice between gains. However, individuals convert to a risk-taking mode if the problem is expressed as a choice between losses. The availability heuristic is another important cognitive factor.[9] This form of perception bias occurs when knowledge or past experiences are brought to mind. Risks tend to get inflated when past events are coupled to the activity or product. In the months immediately following 9/11, the airline industry sustained a drop in passenger numbers because of heightened fears of plane travel.[10]

Other cognitive processes include stereotyping and anchoring bias.[11] In stereotyping individuals draw general conclusions about risks for a class of activities or products based on limited data or experience about a few members of the class that are inferred to be representative. Manufacturers offer guarantees on products in part to counter stereotype bias. If a tiny percentage of a manufacturer's product line is defective, the entire line may be killed because the public infers that the defective products, although small in number, are representative of the entire product line. Anchoring bias occurs when individuals adjust initial estimates of risk to account for new information or experiences. Such adjustments are often inaccurate. The bias occurs because initial estimates of risk anchor any subsequent adjustments by defining the range in which adjustments are made. The 1982 Tylenol® scare discussed in Chapter 1 is an example. When Tylenol was first brought to market it enjoyed an exemplary safety record. The public had a high level of confidence in the product. The 1982 incident in Chicago where some Tylenol tablets were laced with cyanide led to several deaths and a public relations nightmare for the manufacturer Johnson & Johnson.

The manufacturer was able to restore confidence in the product through new tamper-proof packaging and an aggressive public relations campaign. Product integrity was restored because of the impeccable safety record prior to the Chicago incident. Had the incident occurred at the time of first market release, Tylenol would not likely have survived because it would have been regarded by the public as too risky. The initial estimates of the risk (or safety) govern subsequent perceptions and actions.

Risks associated with activities and products also have certain inherent characteristics that impact perception. Some of the more obvious ones are the following: *Voluntariness* of the risk is concerned with whether people get exposed to the risk voluntarily. Driving a car is voluntary but being exposed to air pollution from a chemical plant near one's home is not. Exposure to natural hazards such as radon gas in homes or tornadoes is usually considered as voluntary. The public is more readily accepting of the risk even though consequences (in the case of tornadoes, earthquakes, etc.) may be severe in terns of deaths, injuries, and damage. *Immediacy* has to do with how quickly effects become manifest after exposure. Risks that become evident immediately are viewed as more hazardous than risks whose effects are delayed in part because a latency period leads to uncoupling of the causal agent from the effect. A clear exception to this is cancer. Agents that cause cancer are uniformly feared even though cancers usually take years or decades to appear after the agent exposure. *Knowledge* of the risk also colors perception. If science understands the risk and if people also know something about the risk, it is likely to be more readily accepted. An important determinant of perception is the degree to which individuals have *control* over the risk. Almost every driver feels more comfortable behind the wheel than in the front seat with someone else driving. Accidents seem more imminent when one is in the passenger's seat than behind the wheel. Risks that have high *catastrophic potential* are perceived as more dangerous than risks that affect one or a few individuals at a time. One reason people avoided air travel in the months immediately after 9/11 was the fact that hundreds died in the four plane crashes.[12]

Risks that are perceived to be important in one society or culture may be inconsequential in others. People in Third World countries have a different set of worries than people in wealthier nations because of social, cultural, and economic conditions. In developing countries, people worry about starvation, infectious diseases, malnutrition, and lack of health care or effective public health programs. In affluent countries like the U.S., these problems are essentially nonexistent. Instead, we worry about sophisticated technological risks such as radioactive waste disposal that have very low probabilities of occurrence. Affluent countries have solved major public health problems, including food safety and control of infectious diseases that continue to plague Third World nations. The economies of affluent countries are large enough to control large public health risks. In America we are faced with the risk of obesity, while developing countries are faced with the risks of starvation and malnutrition.

In dealing with public perceptions of risk we need to be careful not to underestimate the significance of perceived risks and the distinction between perceived and real risks. Many experts harbor the traditionalist view that the public is almost always ignorant of the technical issues and that their concerns are emotive rather than rational. Experts think that solving the public misperception problem is a matter of educating the public and to argue that risks are really zero and consequently public concern is irrational.

But understanding how people perceive risks is important and necessary. Technical components of risk (e.g., probability assessments) are important metrics but are often of little concern to the average individual. Nonscientific attributes and dimensions of risk appear to be critical in driving risk perception. Decision makers and regulators need to understand how people think and respond to risks. Without such understanding well-intended policies may be ineffective.[13]

Management Capacity

Prioritization of risks and the decision to manage them depend on the capacity to control the risk. Personal behaviors and events of nature cannot be fully controlled and therefore are somewhat refractory to management. Natural disasters cannot be avoided and personal behaviors by their very nature cannot be controlled very well by outside forces. Interestingly, it is these risks that involve the greatest damage and loss of life. Individual behaviors are either not regulated or regulated to a very limited extent. The government avoids regulating individual behaviors in part because of interference with constitutional rights and individual liberties. Risky activities driven by personal behaviors are subject to regulatory control particularly when the activity is associated with the possibility of collateral injury. Smoking and alcohol consumption are regulated to limit impacts on others (e.g., respiratory illnesses in nonsmokers, a traffic fatality caused by a drunk driver).

Natural disasters represent another class of risks that cannot be readily controlled. Some events like hurricanes and tornadoes occur with high probability. Depending on the circumstances the consequences may be quite severe. Hurricane Katrina (August 29, 2005) may have been the largest hurricane to strike the U.S. gulf coast in history. More than 1,000 persons were killed, tens of thousands more were displaced, and damage has been estimated at $35 billion to $60 billion.[14] There is very little that can be done to prevent or ameliorate the risk, but if given sufficient warning, individuals and communities can take steps to reduce potential damage and loss of life. But even with some advance warning there is no guarantee, as seen with the Katrina disaster, that government authorities will respond in a timely fashion or that individuals will take protective actions.

Volcanic eruptions cannot be influenced by people, but their effects can be slightly. For instance, there are several examples of attempts to divert lava flows before they engulf buildings or towns. One was at Mount Etna when dynamite and bulldozers were used to punch holes in the confining levees on the side of a flow, allowing the molten lava to flow in a different direction, sparing a resort (which was eventually destroyed by a different eruption). Another case was in Iceland where lava threatened to close off a harbor. The lava flow was arrested by spraying cold seawater on its leading edge. There are legal problems with interfering in natural phenomena like lava flows. If such flows cause damage without human interference, that's legally considered an act of God. However, if people try to disrupt or divert a flow and then it causes damage, the agency responsible for the interference may be legally liable for the damage. In the U.S., the principle of sovereign immunity protects government representatives (generally the U.S. Geological Survey) acting in good faith from such liability.

Tsunamis are another example where nothing can be done to prevent the risk from occurring. Damage can be mitigated if protective measures are instituted soon enough. Tsunamis are extraordinary events of nature. They are huge sea waves generated by deep sea earthquakes, landslides, or volcanic activity. Although sea waves may be only 1 to 2 meters high in deep waters, the waves rise to heights of 10 meters or more at the shore. With wavelengths of the order of 150 kilometers, tsunamis travel at jet speeds approaching 800 kilometers per hour. A tsunami generated in the Aleutian Islands would take only 7 hours to reach Hawaii. Communities in areas susceptible to giant sea waves like Hawaii have relocated buildings far enough from shore to diminish the possibility of damage and have set up early warning systems to allow the public time to prepare for the oncoming disaster.[15] The devastating Sumatra tsunami that struck about a dozen countries on the Indian Ocean rim on December 26, 2004, killed more than 200,000 people and displaced 1 million more. This was one of the world's worst natural disasters in decades and has been called the most devastating tsunami in recorded history. The tsunami was caused by a magnitude 9.1 earthquake beneath the Indian Ocean off the western island of Sumatra. Traveling with a velocity of up to 900 kilometers per hour, the tidal wave struck without warning. Coastal towns and villages were left totally unprepared. Had a tsunami early warning system in the Indian Ocean been operative, perhaps thousands of lives could have been saved.[16]

Court Actions

Social regulations are laws to control activities that may negatively impact the environment, health, and safety. Without regulations, firms may not take into account the full social costs of their actions. Government intervention is necessary to ensure that workers have adequate information about workplace health and safety hazards to make fully informed choices and to impose cost controls so that firms do not excessively pollute the environment.[17] Government prioritizes risks by identifying agents requiring regulatory control and by setting exposure limits in the workplace and environment.

When firms or stakeholder groups perceive regulations to be too burdensome, recourse to the legal system is an effective means of obtaining relief. Alternatively, when regulations are either absent or too lax, stakeholders may petition standards-setting organizations in the rulemaking process. Often the science underlying decision making is in question; special-interest groups and other stakeholders can resort to specific legal mechanisms to obtain scientific and technical data.[18]

A number of recent cases illustrates the influence of the courts on risk prioritization through judicial review of agency science and risk assessment. One key question before the courts is the validity of the scientific evidence underlying regulatory decision making. In *Industrial Union Department AFL-CIO v. American Petroleum Institute* (the Benzene Case), the Occupational Safety and Health Administration (OSHA) sought to tighten the benzene limit from 10 parts per million (ppm) to 1 ppm. Industry challenged on the basis that the government had failed to demonstrate "significant risk" at doses below 10 ppm. The court held that "safe is not the equivalent of risk-free" but rather "safe" means "no significant risk." A workplace is only "unsafe" if it poses a "significant risk." A standard must protect against "significant risk," not necessarily all risks. The court remanded for finding by OSHA of a "significant risk" below 10 ppm.[19] In *Chlorine Chemistry Council v. EPA*, the court held that the EPA failed to use its own scientific evidence in regulating chloroform levels in drinking water. The court vacated the EPA's no-threshold-based maximum contaminant level goal (MCLG) of zero because of the EPA's own finding of a threshold dose-response for chloroform.[20]

The courts have also been concerned about the application of cost-benefit analysis in decision making. In *American Trucking Associations v. EPA*, the industry challenged the more stringent National Ambient Air Quality Standards (NAAQS) established by the EPA in 1997 under the Clean Air Act. The industry claimed that the agency failed to consider costs and failed to consider the beneficial health effects of ozone and particulate matter in the air such as screening out solar ultraviolet radiation for the prevention of skin cancers. The court held that the EPA was required to consider both risks and benefits of a pollutant. But the court also supported the agency finding that particulate material "contributes" to endangering the public health based on epidemiological data. Even though the evidence was not proof of causality, the court said the epidemiological data are sufficient to meet the "latest scientific knowledge" provision of the Clean Air Act.[21]

Influence of Stakeholder Groups

Societies, particularly in developed countries, are increasingly interested in more actively participating in the processes of decision making regarding health, safety, and environmental issues. Decision making, regardless of how socially contentious the issue may be, should involve stakeholders to some degree. If stakeholders are included in the decision-making process, they are more likely to buy into any decisions they have played a direct role in. Stakeholders provide valuable perspectives and solutions to risk problems. Further, stakeholder groups can provide invaluable assistance in implementing risk-management decisions if they are made to feel that they are important players in the decision process. The key is to determine the degree of involvement and how early in the decision-making process stakeholders should be included. Decision making may be greatly facilitated if triggers could be identified in advance indicating when stakeholders should be involved and who should be included.

Stakeholders may be defined as individuals or groups that have an interest in or are impacted by a regulatory or policy decision. In theory, stakeholders represent every segment of society and include management and labor groups directly impacted by a particular regulation, consumer groups concerned about product safety, and local communities concerned about health risks associated with local technologies. The value of stakeholder participation in decision making is well recognized. Rulemaking in the U.S. federal agencies requires that they consider input from the public on any proposed rule through a public comment period. Agencies are required to consider public comments in final rulemaking. Optimizing stakeholder involvement in decision making requires careful delineation of the roles of the various stakeholder representatives. Decision makers need to recognize that experts provide important scientific and technical information but their views are neither more nor less important than social and other nontechnical perspectives in the decision-making process. The integrity and credibility of the decision-making process is compromised when stakeholders are made to feel that their perspectives are not valued.

Special-interest groups (defined here as individuals or groups with specific and often self-serving agendas in the decision-making process) do not necessarily represent the public. Special-interest groups have agendas that may not be congruent with views or concerns of the general public. Such groups act with little accountability since their position cannot generally be falsified scientifically. These groups often take an extreme precautionary position that is not in the mainstream of scientific thought and frequently requires proof of the negative to invalidate their position. For example, they argue that since we do not have evidence that agents are without harm at environmental doses, we should assume that they are harmful and take appropriate action to reduce them to levels as close to zero as possible. If they are right, a very small diminution in risk may result that is difficult if not impossible to detect but at enormous costs to society. If they are wrong, a tremendous amount of money will have been expended for no public health or environmental benefit. The real tragedy is that the expended resources could have gone to a more worthy effort to address public health concerns. It should be added that special-interest groups may also raise legitimate issues that would otherwise be of marginal concern to decision makers. Such views are important and deserve serious consideration. For example, community activists may raise legitimate concerns about the impact of the design of a local industrial plant on community health and safety.

Extremist groups often have a disproportionately influential role in risk prioritization and decision making because they are well organized and their message (frequently based on the precautionary principle that it is better to be safe than sorry) has appeal. The influence of such groups persists because the mainstream community fails to counter effectively their views and positions. Extremist groups will weaponize the stakeholder process by taking advantage of Internet and legal remedies such as the Freedom of Information Act (FOIA) and the Federal Advisory Committee Act (FACA) to subvert the work of advisory groups or decision-making processes.[22] Scientists as a group have generally been unwilling to organize and express their views in the political arena. Many scientists are uncomfortable outside of the laboratory and consider the economic, political, and social aspects of their work to be the responsibility of others.[23]

Certain federal and state laws provide a mechanism for public participation in the decision-making process. Activities range from direct participation in the rule-making process by participation in public meetings and submission of documents during public comment periods to input into selection of individuals to committees that advise the government on controversial matters of science and technology. FACA provides the public with the opportunity to review and comment on the appointment of individuals to serve on committees and expert groups that advise government agencies. The establishment of the National Research Council (NRC) BEIR VII committee is an example of how this process works.[24] The Biological Effects of Ionizing Radiation (BEIR) Committees historically have had a high profile because they provide independent estimates of ionizing radiation health risks that are heavily relied upon by federal agencies in the standard setting process. Who gets appointed to the BEIR Committees impacts report findings and indirectly affects standards setting.[25] In 1999, the NRC BEIR VII Committee was assembled and names of the provisional committee members were posted on the Internet for public comment. Committee appointment by the NRC is based on expertise in relevant aspects of the subject area, but philosophical views on the issues are not selection criteria. Antinuclear groups, concerned that their views were not adequately represented and that risk estimates should be increased to support more restrictive dose limits, effectively used the Internet to garner support for a change in the committee composition through FACA.[26] Committees may also censor one of their own members. An NRC panel on endocrine disruptors requested the removal of one member from the committee because of his scientific views and perceived conflicts of interest.[27]

Selection of members to committees is governed to a certain degree by the goal of achieving consensus. Experts are not likely to be asked to serve if their views are known to be extreme such that committee consensus may be compromised. Although members may have differing viewpoints, it is expected that committee reports will represent a consensus view. The desire to achieve a consensus view on a contentious issue like the health effects of low-level ionizing radiation is at cross purposes with the nature of the underlying science. A consensus view may give the false impression of certainty by discounting the legitimacy of other perspectives or other interpretations. In science there can be many legitimate views and interpretations of the same set of data. Consensus views by authoritative bodies may facilitate policy making, but decisions are compromised when perspectives are limited. Consultative bodies such as the National Academies pride themselves on providing quality advice to sponsors. The price of that advice is the desire for a clean consensus document. Although minority reports are not forbidden, they are clearly not encouraged in the report process.[28] Input by experts into the decision-making process should include multiple perspectives presented in a balanced format that includes strengths and weaknesses of differing views. Decision makers need this type of information in order to make sound, scientifically defensible decisions.

REAL RISKS AND REORDERING PRIORITIES

The public health is impacted by many factors, including exposure to biological, chemical, and physical agents, natural events, and technology and other anthropogenic activities. But what are the major risks impacting public health? Although any agent or activity that might affect public health should not be dismissed, which ones are really major and should generate the most concern?

Table 6.3 lists major causes of death in the U.S. and compares mortality data for 1990 and 2000. The data in the table is based on a total of 2.4 million deaths in 2000 and 2.15 million deaths in 1990. Smoking remains the leading cause of death in the U.S., but physical inactivity and poor diet may overtake smoking in the near future. One in six deaths in the year 2000 could be attributed to tobacco use and another one in six is attributable to diseases linked to poor diet and physical inactivity. Together one in three deaths is caused by either tobacco or physical inactivity and poor diet. Among major causes of mortality, diet and poor nutrition have contributed

TABLE 6.3
Leading Causes of Death in the U.S.

Causes of Death	Number (1990)	Percent of All Deaths (1990)	Number (2000)	Percent of All Deaths (2000)
Tobacco	400,000	19	435,000	18
Poor diet and physical inactivity	300,000	14	400,000	17
Alcohol consumption	100,000	5	85,000	4

Source: Mokdad, A.H. et al., Actual causes of death in the United States, 2000, *Journal of the American Medical Association,* 291, 1238, 2004.

the largest percentage increase since 1990. The major causes of death in the U.S. and other First World countries differ considerably from the situation in Third World countries. Socioeconomic conditions, health education, availability of health care, and safety of the water and food supply are major determinants of overall public health.[29]

What is remarkable about this table is the fact that the causes of death are poorly controlled. The agents and activities that are heavily regulated are not included in this list; collectively they contribute only a small number of deaths compared to tobacco, poor diet and lack of physical activity, and alcohol consumption. That does not mean that we should not be concerned about them. On the contrary, any agent that potentially impacts public health should be controlled. This should be done in a way that is consistent with the larger public health problems. If we have the capability to control a risk, we should do so. It does not make much sense to allocate huge resources to control agents that might contribute a few percent to the mortality burden when much larger factors such as smoking and diet involve significantly larger contributions but are left unmanaged.

Prevention and early detection will save more lives from cancer than any other available approaches. It is a general medical and public health principle that preventing disease is preferable to having disease and then treating it. There is no risk of treatment complications if there is no disease to treat. But given that disease is present, the risk of treatment complications may be significant with no guarantee that the disease will be controlled or otherwise acceptably managed. The American Cancer Society estimates that about 60% of the national cancer burden, or 350,000 cancer deaths annually, can be prevented by modifying certain preventable behaviors such as poor nutrition, physical inactivity, obesity, and smoking. In addition, about 1 million skin cancers were diagnosed in 2005 and could have been prevented by the avoidance of excessive sun exposure. Certain cancers related to infectious diseases like hepatitis B and human papillomavirus may be prevented through behavioral changes and implementation of vaccine programs.[30]

These prophylactic measures would appear to be straightforward, but in some cases implementation is problematic. Millions of Americans would benefit from weight-reduction programs. But diet control programs such as the Atkins® and South

Beach™ diets may be out of reach for everyone but the rich. At about $100 per week, most people cannot afford to stick to the specific expensive foods recommended by Atkins or South Beach. Human papillomavirus vaccines for the prevention of cervical cancer have produced encouraging results, but implementing widespread vaccine programs has raised several significant questions.[31]

MONETARY COSTS

In the aggregate, the regulatory program of the U.S. government is a cost-effective enterprise. Regulatory standards for automobile safety, fire safety, clean water and air, clean drinking water, and so forth have clearly resulted in protection of the public health at compliance costs that are exceeded by the public health savings. Table 6.4 presents estimates of costs and benefits of all regulations reviewed by the Office of Management and Budget over the ten-year period from October 1992 to September 2002. These estimates are not a complete accounting of all costs and benefits for all regulations. The table includes only those regulations that generated costs or benefits of at least $100 million and for which a substantial portion of costs and benefits could be quantified.[32]

The total benefits shown in Table 6.4 are roughly three to five times the aggregate costs. A broad range of regulations is included in this table. Many regulations are highly cost-effective but others, such as regulations to control carcinogens, are very expensive and provide little public health benefit. Costs of regulatory control of human carcinogens such as arsenic, benzene, formaldehyde, dioxin, and radionuclide emissions are of the order of several millions of dollars per statistical life saved.[33] Determinants of cost-effectiveness include the magnitude of the managed risks and how well risks and benefits can be quantified.

TABLE 6.4
Estimated Annual Benefits and Costs of Major Federal Rules (October 1992 to September 2002)

Agency	Benefits (2001 dollars)	Costs (2001 dollars)
Agriculture	$3.1–6.2 billion	$1.6 billion
Education	$0.6–0.8 billion	$0.3–0.6 billion
Energy	$4.7 billion	$2.5 billion
Health and Human Services	$8.7–11.7 billion	$3.2–3.3 billion
Housing and Urban Development	$0.5–0.6 billion	$0.8 billion
Labor	$1.8–4.2 billion	$1.0 billion
Transportation	$6.2–9.5 billion	$4.3–6.8 billion
Environmental Protection Agency	$108.9–180.8 billion	$23.9–27.0 billion
TOTAL	$134.5–217.5 billion	$37.7–43.8 billion

Source: Office of Management and Budget, Draft 2003 report to Congress on the costs and benefits of federal regulations; Notice, *Federal Register,* 68, 22, 5492, February 3, 2003.

For carcinogenic agents, risks cannot be quantified reliably and the risk uncertainties are such that quantification of benefits (as measured by the diminution in risk) cannot be easily ascertained. It is difficult to justify the astronomical costs associated with environmental remediation and radioactive waste disposal. The following examples illustrate the magnitude of the problem:

Environmental Cleanup at the Nevada Test Site

Costs of reducing environmental and public health effects of ionizing radiation exposure can be enormous.[34] During the Cold War the U.S. government conducted numerous atmospheric and underground nuclear weapons tests at the Nevada Test Site (NTS). The testing range is heavily contaminated from radioactive fallout. The government estimates that public expenditures will exceed $100 million to clean up radioactively contaminated soil to levels less than 10% of natural background radiation levels. Costs would be significantly reduced if cleanup to levels at or near natural background levels were adopted. Relaxing cleanup standards in this way would not impact public health (Figure 6.1).

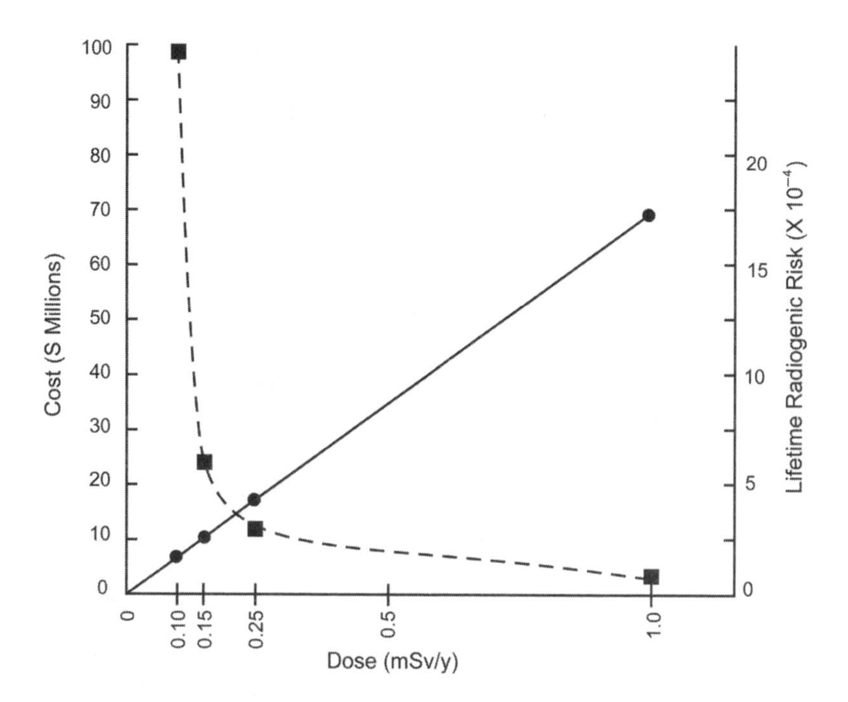

FIGURE 6.1 Economic costs of risk reduction. Cleanup costs (dashed line) rise precipitously despite very small theoretical reductions in lifetime radiogenic risk (solid line) when dose goals are small fractions of natural background levels of 1 mSv per year excluding contributions from radon gas. (From Mossman, K.L., Restructuring nuclear regulations, *Environmental Health Perspectives,* 111, 13, 2003.)

Characterization of Waste Destined for WIPP

The U.S. Department of Energy is responsible for the cleanup of defense-related transuranic waste across the nuclear weapons complex.[35] The Waste Isolation Pilot Plant (WIPP) located near Carlsbad, New Mexico, is the designated repository for this waste. Before shipment to WIPP, waste must be characterized to comply with regulations, including meeting total activity limits and assuring that unallowable items are not included in waste drums. Waste characterization is one of the most costly and time-consuming parts of the national transuranic waste-management program. According to DOE the cost of characterization and certification activities to prepare waste to be shipped to WIPP is estimated to be $3.1 billion. Characterization of a single 55-gallon drum of waste costs about $4,000. A 2004 analysis of the characterization process by the NRC indicates that characterization procedures, including head space gas sampling and visual inspection of drum contents are redundant, very costly, and provide little, if any, impact on repository performance.[36]

RISKS IN PERSPECTIVE

Prioritization of risks is not a static process. New scientific information, changes in legislative agendas and changing perceptions of risk may elevate a previously ignored risk to new heights of public concern. Prior to the 1980s, radon gas was not considered a significant public health hazard, although it was a known cause of lung cancer in uranium and other metal miners exposed to very high concentrations of radon gas in mines. In 1984, scientific evidence indicated that radon levels in houses could approximate those found in mines. In response the EPA spearheaded a national program to control indoor radon levels.

The lack of congruence between perception and reality as reflected by what the public chooses to regulate is an ongoing and troublesome problem. As long as the public continues on a course of inaction toward more strict control of diet, smoking, and alcohol consumption, there will be little progress in reducing the incidence of heart disease, diabetes, and many cancers. Cardiovascular diseases and cancer are the two leading causes of death in the U.S. Reprioritization of risks so that greater control of diet and cigarette and alcohol consumption can be achieved can only be accomplished through political will.[37] Perhaps the most prudent course of action is to gradually tighten existing alcohol and cigarette regulations followed by the introduction of new legislation that controls activities in currently unregulated sectors of society (e.g., cigarette consumption in adults). Some municipalities have already done this by restricting cigarette smoking in public places. The number of smokers fell 11% in New York City from 2002 to 2003 by increasing sales tax, making almost every workplace smoke-free, and assisting smokers to quit by distributing free nicotine patches.[38] The lost tobacco tax revenue is substantially offset by reduced health care costs, increased productivity, and improved public health.

Prioritizing tobacco and other serious public health problems can be facilitated by a strategic shift in how we analyze risks. Typically the public considers risks without regard to other related risks or the benefits of the activity or product generating the risk. Targeting a particular risk may have unintended consequences (see Chapter 5). Shutting down nuclear power plants might ease some concerns about nuclear safety and radioactive waste disposal, but fossil fuel usage would have to be increased to counter the resulting energy shortage. Nuclear power contributes about 20% of the nation's electricity supply. Burning fossil fuels increases atmospheric carbon dioxide, contributes to global warming, and increases U.S. dependence on foreign sources of oil.

A public health approach is needed in assessing and managing small technological risks.[39] Risks should be evaluated in the context of other similar risks. It is impractical to evaluate risks in isolation because risks are not encountered that way in real life. Everyone is exposed to a multitude of risks all the time. Effective individual and public health gains are achieved by identifying the largest risks and allocating resources for their control. It makes little sense to spend huge sums of money to control small technological risks if little is done to control major causes of disease. Accordingly, analyzing a particular risk requires consideration of how that risk impacts the total risk profile of individuals or the community. Risks need to be put into perspective by comparing benefits and risks of alternative activities or products. In this way the small risks that seem to capture our attention can be put into proper context.

NOTES AND REFERENCES

1. Doll, R. and Peto, R. The causes of cancer: Quantitative estimates of avoidable risks of cancer in the United States today, *Journal of the National Cancer Institute,* 66, 1191, 1981.
2. Risk assessment, particularly for cancer-causing agents, is a signal-to-noise problem where reliable measurements of risk depend on discerning a signal through a large background noise (the spontaneous cancer rate). At high doses the signal is usually large enough to observe an effect. However, at small doses relevant in the occupational and environmental settings, the signal may be impossible to detect. Coupling the disease to agent exposure is complicated by the inability to distinguish agent-induced cancers from spontaneous ones (for all but a few cases), and the long latency period that separates exposure and clinical appearance of disease.
3. Slovic, P., Perception of risk, *Science*, 236, 280, 1987.
4. Ibid.
5. Schuler, E., The public has its own view of what is a risk, *Nature,* 425, 343, 25, September 2003.
6. The journalist Gregg Easterbrook has an interesting perspective. He suggests that obsession with small risks may actually be a good thing. It may be a reflection of our growing prosperity. We have the technology to detect smaller and smaller risks and the capability to control them. See Easterbrook, G., *The Progress Paradox: How Life Gets Better While People Feel Worse,* Random House, New York, 2003.
7. For an overview of risk perception see Wilson, R. and Crouch, E.A.C., *Risk-Benefit Analysis*, 2nd edition, Harvard University Press, Cambridge, 2001, and Sunstein, C.R., *Risk and Reason,* Cambridge University Press, Cambridge, 2002.
8. Tversky, A. and Kahneman, D., The framing of decisions and the psychology of choice, *Science,* 211, 453, 1981.
9. Tversky, A. and Kahneman, D., Availability: A heuristic for judging frequencies and probabilities, *Cognitive Psychology,* 5, 207, 1973.

10. Gigerenzer, G., Dread risk, September 11, and fatal traffic accidents, *Psychological Science,* 15, 286, 2004.
11. Tversky, A. and Kahneman, D., Judgment under uncertainty: Heuristics and biases, *Science,* 185, 1124, 1974.
12. Only a few of the major factors that impact risk perception have been discussed Other factors include whether the risk is dreaded and how severe are the possible consequences. See the following for an excellent overview of risk perception: Slovic, P., *The Perception of Risk*, Earthscan, London, 2000; Wilson, R. and Crouch, E.A.C., *Risk-Benefit Analysis,* Harvard University Press, Cambridge, 2001.
13. *Supra* note 3.
14. Barrett, J., Pay now to save later, *Newsweek,* September 26, 2005.
15. Dudley, W.C. and Lee, M., *Tsunami!,* University of Hawaii Press, Honolulu, 1988.
16. For more technical details about the tsunami and the earthquake that caused it, see Marris, E., Inadequate warning system left Asia at the mercy of tsunami, *Nature,* 433, 3, January 6, 2005; Perkins, S., Tsunami disaster, *Science News*, 167, 19, January 8, 2005; Schiermeir, Q., On the trail of destruction, *Nature*, 433, 350, January 27, 2005; Lay, T. The great Sumatra-Andaman earthquake of 26 December 2004, *Science*, 308, 1127, May 20, 2005.
17. Mossman, K.L., Restructuring nuclear regulations, *Environmental Health Perspectives,* 111, 13, 2003.
18. See Data Access Law (P.L. 105–277); Data Quality Law (P.L. 106–554); U.S. Environmental Protection Agency, *A Summary of General Assessment Factors for Evaluating the Quality of Scientific and Technical Information,* EPA 100/B-03/001, 2003.
19. IUD, *AFL-CIO v. API* 448 U.S. 607 (S. Ct. 1980).
20. *Chlorine Chemical Council v. EPA* 206 F.36 1286 (D.C. Cir. 2000). The enforceable standard for chloroform in drinking water is the maximum contaminant limit (MCL). The MCL goal is not a regulatory limit but is used as a guideline for setting the MCL.
21. *American Trucking Associations v. EPA* 175 F.3d 1027, 1055-56 (D.C. Cir. 1999).
22. Freedom of Information Act, 5 U.S.C. 552, as amended; Federal Advisory Committee Act, Pub. L. 92–463, 66 Stat. 770.
23. There are clear exceptions. Many scientific and professional societies maintain relations with lawmakers and regulatory agencies to influence legislation and rulemaking. Individual scientists in academe and elsewhere are creating small technology transfer companies to market their research discoveries.
24. The National Academies consist of the National Academy of Sciences, the National Academy of Engineering, and the Institute of Medicine. The National Research Council is the operating arm of the three academies. The National Academies advise the federal government through written reports by committees of academy members and other experts on a wide range of matters. The BEIR (Biological Effects of Ionizing Radiation) Committees periodically review current scientific literature and advise the federal government on the effects of low-level ionizing radiation. The magnitude of the health effects of low-level radiation is highly controversial and the scientific community is split as to whether low levels of ionizing effects increase cancer, decrease cancer (i.e., hormesis), or have no effect. The seventh and most recent BEIR report was published in 2005. See National Research Council, *Health Risks from Exposure to Low Levels of Ionizing Radiation,* BEIR VII Report, National Academies Press, Washington, DC, 2005.
25. The National Research Council BEIR III report published in 1980 contains several minority reports reflecting views of committee members not entirely consistent with the main report. The spectrum of views reflected preferences for different theories to estimate risk. See National Research Council, Committee on the Biological Effects

of Ionizing Radiation (BEIR), *The Effects on Populations of Exposure to Low Levels of Ionizing Radiation,* BEIR III Report, National Academy Press, Washington, DC, 1980.

26. One letter, circulated via the Internet and signed by some 70 antinuclear groups, resulted in the removal of three provisional members because of their "alleged" pro-industry views. The antinuclear groups argued that the provisional committee was not philosophically balanced in its views on low-level radiation risks. The National Academies responded by replacing the individuals.
27. Safe, S., Endocrine disruptors. In Gough, M. (Ed.), *Politicizing Science: The Alchemy of Policymaking,* The Hoover Institution Press, Stanford, CA, 91, 2003.
28. *Supra* note 25.
29. Mokdad, A.H. et al., Actual causes of death in the United States, 2000, *Journal of the American Medical Association,* 291, 1238, 2004.
30. For a detailed discussion of U.S. cancer statistics see American Cancer Society, *Cancer Facts and Figures 2005*, American Cancer Society, Inc., Atlanta, GA, 2005.
31. Schiller, J.T. and Davies, P., Delivering on the promise: HPV vaccines and cervical cancer, *Nature Reviews Microbiology*, 2, 343, 2004.
32. Office of Management and Budget, Draft 2003 report to Congress on the costs and benefits of Federal regulations; Notice, *Federal Register,* 68, 22, 5492, February 3, 2003.
33. Tengs, T.O. et al., Five hundred life saving interventions and their cost effectiveness, *Risk Analysis*, 15, 369, 1995.
34. Mossman, K.L., Restructuring nuclear regulations, *Environmental Health Perspectives,* 111, 13, 2003.
35. National Research Council Committee on Optimizing the Characterization and Transportation of Transuranic Waste Destined for the Waste Isolation Pilot Plant, *Improving the Characterization Program for Contact-Handled Transuranic Waste Bound for the Waste Isolation Pilot Plant,* National Academy Press, Washington, DC, 2004.
36. Ibid.
37. Attempts to completely eliminate cigarettes and alcohol will probably be futile in much the same way that prohibition of alcohol failed in the 1920s and 1930s. Any legislation that impacts availability of cigarettes and alcohol must recognize that although the greater public good is served by such action, some businesses will be seriously impacted. Among other things, federal legislation must incorporate subsidies and tax write-offs for businesses that stand to be impacted by such legislation through lost revenues.
38. New York City Department of Health and Mental Hygiene, Office of Communications, *New York City's Smoking Rate Declines Rapidly from 2002 to 2003, the Most Significant One-Year Drop Ever Recorded*, May 12, 2004.
39. The 1997 Presidential/Congressional Commission on Risk Assessment and Risk Management also discusses evaluation of risk in a public health context but in a more limited way by focusing on the comparisons of specific risks related to a proposed risk-management action. See Presidential/Congressional Commission on Risk Assessment and Risk Management, *Risk Assessment and Risk Management in Regulatory Decision-Making*, Final Report Volumes 1 and 2, 1997. http://www.riskworld.com/ riskcommission/Default.html (accessed March 2006).

7 Avoiding Risk

Exposure to agents that cause cancer and other diseases occurs in everyday life and particularly in certain workplace settings. In an urban environment everyone is exposed to smog and other air pollutants in addition to natural background radiation and solar ultraviolet light. An estimated 50 million people currently smoke in the U.S. Millions more experience the irritating effects of secondhand (environmental tobacco) smoke.[1] Some agents we are exposed to are known or suspected human carcinogens. How should the collective impact of these agents be evaluated? Analyzing the health impact of a single agent without regard to the presence of competing risks is inappropriate because perspectives on isolated risks can be distorted easily. Agents may interact in ways that enhance or diminish risk. The logical approach to dealing with multiagent exposures is to express health impacts for each agent in terms of a common currency — risk. Ideally health protection frameworks for carcinogens should be coherent. In principle a set of common assessment and management concepts and tools can be developed by using a risk-based system that may be applied to a broad array of agents.

Over the past 50 years, radiation protection has evolved into a risk-based system with the goal of establishing a coherent framework of protection. In this way the health detriment associated with internal or external radiation exposures involving different ionizing radiation types (e.g., x-rays, neutrons, alpha particles) can be reduced to a single number for risk-management purposes. The International Commission on Radiological Protection (ICRP) introduced a risk-based system in 1977 as a solution to the problem of combining doses from different radiation sources.[2] However, the system has created more problems than it has solved. Management of chemical carcinogens is similarly based on a risk framework. Risk is the coin of the realm, and decision making is facilitated by reducing health effects from multiple agents to single risk numbers.

In previous chapters arguments are made that the high degree of uncertainty associated with very small risks makes risk-management decisions difficult. Further most people can't put risks into perspective because they do not comprehend small probabilities very well. There is little personal experience with low-probability events that characterize the vast majority of carcinogen exposures (cigarette smoking is an obvious exception), so it is necessary to use statistical probabilities for expression. The public does not think effectively in statistical terms. Risk misconceptions can lead to confusion and public fear and may impair communication.

In this chapter the case is made to use dose as the basis for assessment and management of carcinogens instead of risk. The underlying principle is that decision making should be based on what is known, and uncertainties should be minimized wherever possible. Risk estimates are more uncertain than dose estimates because risk is derived from dose. However, dose estimates can also be uncertain depending on complexity of dose models and confidence in input parameters.

THE CASE AGAINST RISK

Several simplifying assumptions are necessary to make a risk-based system workable and practical. First, health consequences from exposure to different carcinogenic agents are assumed to be the same. If risks are not the same, they can neither be combined nor compared in a meaningful way. Second, risks are assumed to be independent and can be added. If agents interact or have overlapping mechanisms of action then the combination of risks is nonlinear. Third, dose is assumed to be a surrogate for risk since small risks cannot be measured directly but dose can be. For very small risks these assumptions have proven to be untenable. None of these assumptions is sufficiently robust to support risk as a basis for decision making.

DIFFERENT RISKS

Agent-specific health risks are often not comparable. In a risk-based system it is assumed that a 1:1,000 risk from agent A has the same meaning as a 1:1,000 risk from agent B. But agents frequently cause different cancers with different management challenges and outcomes. Lung cancer, for example, has a mortality rate approaching 90%, whereas thyroid cancer is highly curable and has a mortality rate of less than 10%. If agents cause different diseases, their risks cannot be combined in any meaningful way.

The problem extends to risk comparisons. The U.S. Environmental Protection Agency (EPA) has used comparisons with cigarette smoking to put radon risks in perspective. Living in a house with a radon concentration of 150 Bq/m^3 (the EPA action level) is equivalent to a lung cancer risk from smoking half a pack of cigarettes daily.[3] But such comparisons are not very helpful because smokers may interpret smoking risks differently than never-smokers. Unlike residential radon risks, health effects of smoking have been well characterized based on direct observations of lung cancer mortality. Although lung cancer is the only significant health outcome from radon exposure, it is only one of several serious health effects associated with smoking.

To compare or to combine risks would require application of a conversion factor. This is equivalent to using exchange rates to convert foreign currencies. There is no straight forward way to "convert" one cancer to another for the purposes of summing risks.[4]

The problem is significant in radiological protection because different cancers are induced depending on the part of the body irradiated. The ICRP recognized this problem and proposed a protection system based on whole-body equivalent doses.[5] Doses from external exposures to parts of the body or from internalized radionuclides that deposit or concentrate in specific tissues are converted to whole-body equivalent doses. In theory, whole-body equivalent doses produce the same spectrum of cancers and can be combined to calculate single risks. This assumes that risks are equivalent whether the whole body is irradiated homogeneously or nonuniformly.[6] In this system, appropriate tissue-weighting factors are used to account for differences in tissue-specific risks when the body is irradiated nonuniformly. The tissue-weighting factor is the ratio of the tissue-specific risk to the whole-body risk. Using tissue-weighting factors, doses from internally deposited radionuclides may be added to doses from external sources to estimate the total dose from all sources to an exposed individual. The U.S. Nuclear Regulatory Commission (U.S. NRC) used this methodology as a basis for the 1991 revision of its standards for protection against radiation.[7]

Although this risk-based system allows for the calculation of a single dose value for comparison with limits, there are serious problems with its utility in radiation protection. First, there is uncertainty in the values of tissue-weighting factors. ICRP recognized these uncertainties but nevertheless assigned single values to these factors to facilitate dose calculations. These factors are based on risk estimates derived from populations exposed to high doses (> 200 mSv) delivered at high dose rates (perhaps 100 mSv/h and higher) but are applied to occupational situations that involve low doses (< 10 mSv) delivered at low dose rate (perhaps 1–5 mSv/y).[8] Age, gender, other host factors, and the shape of the dose-response curve are known to modify risk significantly. At doses near natural background radiation levels (approximating many occupational exposure situations), the range of uncertainty in the lifetime radiogenic cancer mortality risk is large and the lower bound of uncertainty includes zero.[9] Although assigning specific values to tissue-weighting factors facilitates calculations, it is overly simplistic and fails to account for the influence of known risk determinants.

Second, and more importantly, risks are not necessarily equal when comparing whole-body and specific-tissue exposure. This is an important problem when internally deposited radionuclides concentrate in particular tissues. Radiogenic lung cancer from radon exposure is an example where the calculation of a whole-body equivalent risk is inappropriate. Lung cancer is the only known health effect of radon gas exposure.[10] Extrapolation of the lung cancer mortality risk to a whole-body risk is not consistent with current epidemiological understanding of radiological health effects because radon does not result in excess cancers in other tissues and organs of the body. Risk is limited only to the lung; there is no equivalent whole-body dose that results in the same risk.

The National Council on Radiation Protection and Measurements (NCRP) has used the ICRP risk-based methodology to determine that radon gas accounts for about half of the total natural background radiation dose based on a comparison of equivalent whole-body doses from radon and from cosmic radiation and terrestrial radionuclides.[11] Since radon gas has been linked only to lung cancer, it is inappropriate to calculate equivalent whole-body doses. Accordingly radon gas and its progeny should be considered separately from other natural background radiation sources. In occupational settings involving internal exposures to radionuclides that result in uptake in specific tissues (e.g., iodine uptake by the thyroid gland) it is also not meaningful to calculate a whole-body dose.

Agent–Agent Interactions

To arrive at a total health risk from multiagent exposures, it is assumed that the agents act independently. Summing risks assumes that no synergy or other type of interaction occurs. Independency means that agents produce effects by separate mechanisms. If there is mechanistic overlap (for example, common molecular targets or damage pathways), then the total risk would be expected to be greater than or less than the sum of the individual risks. Whether one agent amplifies or diminishes the effects of another agent depends on the mechanisms of action for each agent.

There is little evidence that the assumption of agent independence is valid. Risks are not likely to be additive because carcinogenic mechanisms in specific tissues probably have some common features. For almost all multiagent exposures, interactions are poorly understood. Even for well-known agent combinations like smoking and radon exposure, the interactions appear very complex, and risks cannot be simply added.[12]

The idea behind a risk-based system is to create a common currency whereby health detriments from exposure to various agents can be compared and combined. If the assumption of independence is not valid, adding risks may seriously overestimate or underestimate total risk (depending on whether one agent diminishes or enhances the effect of another).

Dose as a Surrogate for Risk

Uncertainties in cancer risks reflect substantial difficulties in measuring a small effect in the presence of a large natural burden of disease. The presence of confounding factors further complicates assessment of agent-specific risks. Uncertainties lead to misinformation and misperceptions about small risks. Dose is used as a surrogate for risk because it can be measured reliably to very low levels and at levels orders of magnitude lower than risks. Two important assumptions are made when dose is used as a measure of risk. First, risk coefficients (risk per unit dose) are constant over the range of dose of interest. Second, radiogenic health effects occur as a result of direct damage to cells that are irradiated and absorb energy.

The first assumption addresses the problem of converting measured dose to risk. In cancer risk assessment the linear no-threshold theory (LNT) is the basis for this conversion. LNT is simple and straightforward and assumes that the risk coefficient (i.e., the health risk per unit dose) is constant with dose. The slope of the LNT dose response is the risk coefficient, and when multiplied by the given dose provides an estimate of the risk. However, radiogenic cancers have different dose-response characteristics. Bone cancers have a well-defined threshold dose; some types of leukemia have sublinear dose responses. Breast and thyroid cancer follow an LNT dose response. When dose responses are nonlinear the risk coefficients change with dose, and application of LNT theory may seriously overestimate certain cancer risks.

The second assumption addresses the mechanistic relationship between dose and risk. Historically, radiation effects have been assumed to be the direct consequence of energy deposition in irradiated cells. If the assumption is correct, "dose" (a radiological quantity that measures energy absorbed) is an appropriate way to measure risk. However, recent radiobiology studies seriously challenge this assumption and call into question the validity of dose as a surrogate for risk. Nontargeted effects of radiation including bystander effects and genomic instability challenge the idea that only cells that are "hit" by radiation and absorb energy are responsible for subsequent radiobiological effects.

The bystander effect refers to the capacity of cells affected directly by radiation to transfer biological responses to other cells not directly targeted by radiation. Responses in nontargeted cells can be beneficial or detrimental. The term *adaptive response* refers to a biological response whereby the exposure of cells to a low dose of radiation induces mechanisms that protect the cell against the detrimental effects of other events or agents, including spontaneous events or subsequent radiation exposure. Intercellular communication (i.e., cell signaling) implies that the damaged cell elaborates one or more diffusible chemical signals that affect neighboring nontargeted cells.

Genomic instability refers to the acquisition of genetic damage in cells derived from cells damaged directly by exposure to carcinogens, such as ionizing radiation. Instability manifests itself in many ways, including changes in chromosome numbers, changes in chromosome structure, and gene mutations and amplification. Bystander effects are the most likely drivers of genomic instability. Effects can be observed at delayed times after irradiation and manifests in the progeny of exposed cells multiple generations after the initial insult. Genomic instability is important in the cancer initia tion process and appears to be a critical element in cancer progression and metastasis.

Nontargeted delayed effects suggest that risk is fundamentally different at high and low doses, and simple dose extrapolations to predict risk do not adequately account for the complexity of effects in the low-dose range. As dose is reduced, nontargeted effects become increasingly more important. At about 5 mSv approximately one "hit" occurs, on average, in a population of exposed cells. Assuming cells are hit by radiation in a random fashion and that the statistical distribution of hits follows a Poisson distribution, about 37% of cells will receive no hits if the average number of hits is 1 per cell (clearly some cells will sustain multiple hits). If the dose is reduced tenfold (i.e., the average dose is about 0.5 mSv and the average number of hits is 0.1 per cell), the percentage of cells that are not hit increases to 90%; if the dose is reduced again tenfold the percentage of cells not hit increases to 99%. Cells that are hit sustain the same damage, but more cells avoid damage altogether as dose is decreased. At high doses (e.g., doses greater than 1,000 mSv) essentially all cells sustain at least one hit; very few, if any, cells escape direct damage.[13]

Most studies on nontargeted effects have been conducted *in vitro* with little evidence that such effects have generalized importance in intact tissues and organs. Nontargeted effects should be considered in developing models of carcinogenesis particularly in the low-dose range where nontargeted effects predominate. Current risk estimates likely include bystander influences because risk is an expression of radiobiological responses of tissues and organs rather than of individual cells. Nevertheless, the validity of absorbed dose as the relevant radiobiological quantity to measure risk must be called into question because the sphere of radiogenic effects is larger than the site of energy absorption. Nontargeted effects present new challenges to evaluating cancer risks but the importance of nontargeted effects to carcinogenic risk is not yet clear.

THE CASE FOR DOSE

A framework for protection based on dose has decided advantages. Dose can be measured directly.[14] Measurement techniques are now so sensitive that agents (including ionizing radiation) can be detected at orders of magnitude lower than necessary to produce health effects. The ability to detect very small doses has created a public relations nightmare for many industries. The public assumes that any dose produces some harm and that if the agent is present even in the smallest amount, it must be harmful. Although simply detecting the presence of an agent is not evidence of health risk, it is important to measure doses as low as possible in order to fully comprehend the nature of environmental or workplace exposures. A complete understanding of the distribution of doses in a defined population is necessary in order to plan management strategies.

"Dose" is a concept the public readily understands because of experiences in everyday life. Almost everyone knows what is meant by a "heavy" smoker or drinker by the number of cigarettes or alcoholic beverages consumed in a given time period. Prescription and over-the-counter medications are usually dispensed as pills (i.e., dose units). Some people cut pills in half (i.e., reduce the dose) to get the desired effect if they are sensitive to the medication. The public also has a good sense of what is meant by dose in the context of time and space. Dose rate is a major determinant of health effects; a dose of an agent administered acutely is more biologically effective than chronic administration. Almost everyone knows that six beers in an hour are likely to make you drunk, but the same six beers consumed at the rate of one per day won't have much effect. People also know that route of administration dictates effectiveness. Eating cigarettes won't satisfy a smoker. Addicts understand that for some drugs intravenous injection is more effective than inhalation for a drug-induced high.

A dose-based system avoids the requirement of using a predictive theory to translate dose into risk. Alternative biologically plausible theories predict very different risks at low doses, and predicted risks are highly uncertain because of the large-dose extrapolations required.

Uncertainties in risk are larger than uncertainties in dose because risk is derived from dose. But doses themselves can have wide ranges of uncertainty depending on how they were determined. Doses that were measured directly (e.g., personal dosimeter) typically have small uncertainties. The level of uncertainty will depend on the nature of the source term and detector characteristics. Doses derived from complex models (e.g., models to estimate radiation dose to specific tissues or organs from internal deposition of radionuclides) can be quite complex and uncertain depending on the validity of modeling assumptions and knowledge of input parameters. Dose reconstructions are particularly problematic. In this process one estimates radiation doses received by individuals or populations at some time in the past. Dose reconstructions have been conducted for atomic veterans who participated in various activities during atmospheric testing of nuclear weapons. Incomplete records or questionable information (obtained from memory of events that occurred several decades previously) can lead to highly uncertain estimates.

In a dose-based system measured or calculated doses are compared to regulatory dose limits. It is generally assumed that dose limits are linked to quantitative estimates of cancer risk and that increased risk leads to more restrictive limits. But it is unclear that this is the case (Figure 7.1).[15] Economic, political, and technologic considerations are principal drivers in the standards-setting process and are reflected in safety factors used to keep limits well below observable risk levels. Decision makers are faced with the difficult task of setting limits that protect workers and the public but are not so restrictive that compliance is neither technologically feasible nor economically crippling.

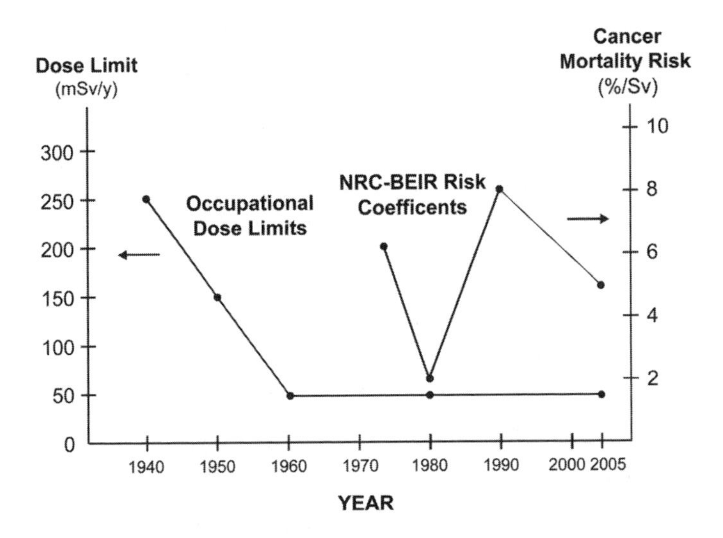

FIGURE 7.1 Dose limits are unrelated to cancer risks. Dose limits are set at levels far below known risks by using safety factors that include economic, political, and social considerations. Limits are rarely if ever relaxed once established. They either remain the same or are made more restrictive. (Data from National Research Council, *The Effects on Populations of Exposure to Low Levels of Ionizing Radiation,* BEIR I Report, National Academy of Sciences, National Research Council, Washington, DC, 1972; National Research Council, *The Effects on Populations of Exposure to Low Levels of Ionizing Radiation: 1980,* BEIR III Report, National Academy Press, Washington, DC, 1980; National Research Council, *Health Effects of Exposure to Low Levels of Ionizing Radiation,* BEIR V Report, National Academy Press, Washington, DC, 1990; National Research Council, *Health Risks from Exposure to Low Levels of Ionizing Radiation,* BEIR VII Report, National Academies Press, Washington, DC, 2005.)

Occupational dose limits have decreased over time since the first limits were established early in the 20th century but have remained essentially constant since 1960.[16] The changes in limits reflect evolving understanding of important health effects following radiation exposure. Before and during the Manhattan project (1942 to 1945), limits were based on tolerance doses for deterministic effects like erythema (i.e., skin reddening). Deterministic effects require that a dose threshold be exceeded. In the years following World War II, tolerance doses gave way to the permissible dose concept, based on the idea that genetic effects and cancer were the important health consequences. These effects occurred without a dose threshold. Since 1960 occupational dose limits have been based on cancer as the major health effect of concern at low doses.

Cancer risks were not known very well in 1960. The National Research Council (NRC) Biological Effects of Ionizing Radiation (BEIR) Committees first estimated risks with the BEIR I report released in 1972. The principal database used to estimate risk has been the Japanese survivors of the atomic bombings. Differences in BEIR risk estimates since 1972 reflect improved epidemiological data, improved atomic bomb dosimetry, and use of different risk-projection models and methods (including temporal projection and population transfer). It is interesting that the risk estimates suggested by the first BEIR Committee in 1972 are not very different from the estimates proffered by the BEIR VII Committee in 2005 (Figure 7.1). But uncertainties in these estimates have been reduced over time because of substantially more epidemiological data and a clearer understanding of the processes of carcinogenesis.

A dose-based system provides limited information about health risks and avoidable health detriment. Doses encountered in environmental and occupational settings are well below observable risk levels since regulatory limits are established with ample safety margins. Measured reductions in dose are assumed to translate into health benefit although the magnitude of any diminution in risk is determined theoretically and cannot be measured. In a dose-based system, public health benefit must be thought of in terms of dose reduction. The goal is to reduce dose, taking on faith that dose reduction leads to a concomitant reduction in risk. Current protection frameworks are "faith-based" because differences in risks before and after an as low as reasonably achievable (ALARA) or other management process cannot be measured.

A serious drawback to a dose-based system is the requirement that agents be considered separately since doses from different agents cannot be combined or considered collectively. Sensitivities to agents vary, and different health effects may occur. Accordingly, agent-specific doses must be evaluated individually and compared to agent-specific limits. This complicates the regulatory framework because of the necessity to construct a complex array of dose limits tailored to individual agents under differing exposure scenarios.

A DOSE-BASED SYSTEM OF PROTECTION

The goal of any system of protection is a framework that is simple, scientifically defensible, understandable to decision makers and the public, and flexible enough to have broad applicability. However, flexibility has its limitations. It is unreasonable to conceive of a common framework of protection for chemical and physical carcinogens. Chemicals and radiation interact with cells and tissues in different ways, and the philosophical approaches to management are entirely different.

A dose-based system may be characterized by three independent dose reference points. The *regulatory dose limit* establishes the ceiling above which occupational or public doses are not permitted. The *natural background level* establishes a lower bound as defined by levels of the agent that occur naturally. The *acceptable dose* is between these extremes and reflects the level of dose for which no further dose management is necessary or required.

What distinguishes dose management systems are the values of the dose reference points for specific chemical and physical carcinogens. Values for the regulatory limit and natural background levels are carcinogen-specific; the acceptable dose for a specific carcinogen may vary depending on the nature of the source term and local technical, social, and economic considerations.

Regulatory Dose Limit

In radiological protection dose limits establish a regulatory ceiling. The overarching philosophy in radiation protection is that doses are kept ALARA below the limit with due consideration given to social and economic constraints. Authoritative bodies such as the ICRP and the NCRP and standards-setting organizations should consider recommending or adopting a dose-based system of protection whereby types of radiation and exposure scenarios (internal versus external dose) are considered separately. To determine compliance with dose limits, dose proportions are calculated by dividing the measured or calculated dose by the relevant dose limit (the dose proportion is dimensionless, since the measured or calculated dose and the dose limit are in the same units). Proportions less than unity would be considered compliant. In the case of multiple agents from a single controllable source, dose proportions for each agent are summed to determine if the sum is less than unity.[17]

A dose-based system requires that standards-setting organizations establish an array of dose limits that recognize differences in radiation types, exposure scenarios, and tissue radiosensitivity. Separate dose limits would be required for the whole body and for specific tissues and organs. Construction of a large matrix of regulatory limits is administratively burdensome but does not present any insurmountable practical problems. Regulatory decision makers need to consider carefully what weighting factors should be used to address radiobiological effectiveness of different radiation types, exposure scenarios, and radiosensitivity differences in tissues and organs. Once the dose limit matrix has been created, practical implementation is relatively straightforward by comparing individual or population doses directly to relevant limits and calculating dose proportions.

Absorbed dose is the preferred quantity because it is most closely related to health effects. However, absorbed dose is not a practical quantity in many instances, particularly for internally deposited radionuclides. Other quantities such as activity concentration may be more practical. Calculations of absorbed dose to the lung from radon gas exposure are problematic in part because of the highly nonuniform deposition pattern of radon progeny in the lung. It is more practical to express radon in terms of activity concentration in air (Bq/m^3). Activity concentration is measured directly. Variability and uncertainty in measurements have been well characterized. Regulatory limits must be expressed in the same quantities and units as measured or calculated exposures or doses to calculate dose proportions.

Natural Background

The natural background level establishes a lower bound for dose reduction. Reducing doses to levels below background necessitates removing some or all sources of naturally occurring terrestrial radionuclides in addition to all exposures from the anthropogenic sources being managed. Such efforts are usually enormously expensive resulting in little or no environmental or public health benefit.

A key recommendation of a 1999 Airlie House International Conference was that reference to natural background radiation should be included in policy discussions on the regulation of radiation sources delivering low-level radiation.[18] Natural background radiation is the largest source of human radiation exposure, and it has been well characterized. Radiation levels vary by geographic location and altitude and have been measured with a high level of accuracy. Natural background radiation levels in areas such as Ramsar, Iran, are so high that annual radiation levels exceed occupational exposure limits.[19] Background varies by about a factor of two in the U.S., with the highest levels in the Rocky Mountains and the lowest readings in the mid-Atlantic states. Excluding contributions from radon gas, the average natural background radiation level is about 1 mSv per year. Epidemiological studies of health effects in populations living in high background radiation areas show no increase in public health effects that may be attributed to radiation exposure.[20]

Many human carcinogens occur naturally, including carcinogens of environmental concern such as dioxin and polychlorinated biphenyls (PCBs).[21] A parallel system of dose-based protection can be established for chemical carcinogens for which natural sources can be identified and characterized (Table 7.1). Many chemical carcinogens are found in the natural environment either as a natural component of

TABLE 7.1
Natural Sources of Selected Known and Suspected Human Carcinogens

Carcinogen	Source	Route of Exposure	Background Levels in U.S.
Ionizing radiation	Cosmic radiation, terrestrial radionuclides, excluding radon	External and internal exposure (ingestion)	~1 mSv/year
Ionizing radiation	Radon	Internal exposure (inhalation)	20 Bq/m^3 (outdoor air); 50 Bq/m^3 (indoor air)
Ultraviolet radiation	Solar UV	External exposure	Considerable variation depending on time of day, altitude, latitude, weather conditions, ozone
Tobacco	Tobacco products	Inhalation; direct contact with oral mucosa	~50 million adults are active smokers; average consumption is half a pack per day
Alcohol	Beer, wine, other products containing ethyl alcohol	Ingestion (diet)	20 ml (serving of beer) 30 ml (serving of wine)
Ethylene dibromide (EDB)	Grains and grain products	Ingestion (diet)	~0.4 μg/day
Polychlorinated biphenyls (PCBS)	Contaminants in food	Ingestion (diet)	0.2 μg/day
Acrylamide	Dry cereals, french fries, potato chips	Ingestion (diet)	10–2,500 parts per billion

Sources: National Council on Radiation Protection and Measurements, *Natural Background Radiation,* NCRP Report 93, NCRP, Bethesda, MD, 1987, for ionizing radiation; Caldwell, M.M., Flint, S.D., and Searles, P.S., Spectral balance and UV-B sensitivity of soybean: a field experiment, *Plant, Cell, and Environment*, 17, 267, 1994, for UV; American Cancer Society, *Cancer Facts and Figures 2005*, American Cancer Society, Inc., Atlanta, 2005, for tobacco; Ames, B.N., Magaw, R., and Gold, L.S., Ranking possible carcinogen hazards, *Science,* 236, 271, 1987, for alcohol, ethylene dibromide and polychlorinated biphenyls; U.S. Food and Drug Administration, *Exploratory Data on Acrylamide in Foods*, February 2003, for update for acrylamide.

the environment, as pollutants in air or water, or as residues or contaminants in foods. Natural levels of these carcinogens are not known to cause elevated rates of cancer in human populations. Yet anthropogenic sources of these agents are regulated to levels smaller than what occurs naturally.

Natural background sources can serve as a useful benchmark for placing doses from anthropogenic sources of chemical carcinogens and ionizing radiation into perspective. But doing so presents challenges. Dose comparisons might not be very helpful in facilitating public understanding of sources of exposure and effects. Although natural background radiation is well characterized, the multiple sources of background can be confusing to the public. Cosmic radiation, external terrestrial radiation, and radiation from internally deposited radionuclides are not understood very well by the public. The public perceives anthropogenic and naturally occurring radiation sources differently even though health risks are the same. Technology-derived radiation is considered more toxic than naturally occurring radiation. The public is apathetic toward health risks from naturally occurring radon gas in homes but expresses heightened concerns for health and environmental risks associated with sites for the disposal of radioactive waste that involve doses representing tiny fractions of annual background radiation levels.

ACCEPTABLE DOSE

In the context of a dose-based system of protection, acceptable dose is the dose that is considered safe. The problem is: safe for whom? The value of the acceptable dose is bounded by the regulatory limit as a maximum and natural background levels as a minimum. The background level is by nature acceptable because it is irreducible. Anthropogenic dose acceptability is a moving target. A dose deemed acceptable to one individual or group may be unacceptable to another because of differences in perception of derived technological benefits and risks. People are likely to accept higher doses (and risks) if there are tangible benefits (e.g., competitive salaries, employment benefits, job security).

Acceptable dose may be operationally defined in the context of an ALARA process. Since ALARA involves balancing the benefits of dose reduction with economic and social costs, the residual dose following an ALARA process may be deemed acceptable. If the residual dose is not acceptable, additional costs are warranted to reduce dose to an acceptable level. Clearly economic and social costs applied to the ALARA process vary locally, and the resulting acceptable (residual) doses will be reflected in the availability of resources to manage dose.

Consideration has also been given to basing acceptable dose on the standard deviation of the average annual natural background radiation level (about 0.2 mSv per year). The health detriment of a small additional anthropogenic dose would be undetectable and acceptable.[22] However, a 0.20 mSv per year standard is unnecessarily restrictive. It is about three orders of magnitude below doses associated with statistically significant radiogenic health risks in adult populations. Others have proposed a less restrictive approach by recommending the annual natural background level rather than the standard deviation of the average annual level.[23] The Health Physics Society has defined acceptable dose as 1 mSv per year above natural background radiation based on current understanding of radiogenic risks from epidemiological studies of cancer in exposed human populations. At this dose level radiogenic health risks are either zero or too small to be measured reliably. If risks exist, they are not of individual or public health concern.[24]

Ultimately setting an acceptable dose level is an economic, political, and social process that should be transparent to all interested and impacted stakeholders. Current scientific understanding of risk should not be ignored. Although epidemiological evidence of risk at natural background levels is lacking, LNT proponents and other conservative thinkers argue that the risk of exposure to varying levels of background radiation is not benign because current epidemiological methods have limits of detection. Accordingly, results are simply inconclusive because of low statistical power and the presence of confounders.

MANAGEMENT DECISIONS BASED ON DOSE PROPORTION

The operational quantity in a dose-based system of protection is the dose proportion. The dose proportion is used in two ways depending on the reference quantity selected. When used to determine compliance with regulatory limits, the appropriate dose limits are the reference.[25]

Dose proportion can also be used to trigger radiation protective actions. A generic decision framework is shown in Table 7.2 for ionizing radiation using natural background radiation levels as the reference. Dose proportions are expressed as decade multiples of natural background radiation levels (assumed to be 1 mSv per year excluding contribution from radon). Protective actions and risk-management decisions are determined by the magnitude of the dose proportion.

Dose proportions greater than 100 warrant serious and immediate concern. In this open-ended dose range, public and occupational dose limits are exceeded and deterministic effects are possible (at doses in excess of about 500 mSv). Risks of delayed effects, especially cancer, are also significant. The nature of protective actions and dose management depends on the exposure situation. In a radiation accident involving release of radioactive materials, evacuation and relocation of

TABLE 7.2
Dose Proportions in Radiation Protection Decision Making

Dose proportion	Likely effects	Concerns
>100	Deterministic effects, mortality possible at high doses; significant risk of cancer	Serious; occupational dose limits exceeded
> 10	Low risk of cancer	High; public dose limits exceeded; occupational doses may be exceeded
>1	Theoretical risk of cancer	Low to moderate
1	Theoretical risk of cancer	Trivial to low

nearby communities may be necessary and require coordination and implementation in a short period of time.[26] Dose proportions greater than 10 but less than 100 warrant a high level of concern because doses exceed the public dose limit. They may exceed or at least represent a significant percentage of the occupational dose limit. Risks of cancer are small and are difficult to measure, but deterministic effects are not likely to occur. Nevertheless, doses should be reduced in accordance with an ALARA philosophy. Dose proportions up to 10 pose low to moderate concerns. In this range public dose limits are exceeded. Cancer risks have not been observed directly at these low doses but are assumed to be nonzero. Consistent with safe practices, doses should be reduced in accordance with the ALARA philosophy. Doses at or below natural background radiation levels (i.e., proportions less than or equal to 1) pose no special concerns and should not warrant any specific protective actions.

Use of dose proportions does not assume *a priori* knowledge of specific health risks. What is implied is that health risks increase monotonically with dose proportion. Dose proportions are an effective way of communicating health risks without resorting to probabilistic risk expressions. It is important to clearly establish the reference dose in the dose proportion. Unless this is clearly defined, the dose proportion is confusing and subject to misinterpretation.

The dose proportion provides a rational framework for radiation protection decisions. Dose limits may be expressed as dose proportions to establish a regulatory perspective. In Table 7.2 the dose proportion would be 50 at the occupational dose limit and 1 at the public dose limit assuming a reference background radiation level of 1 mSv per year. Other dose proportions (e.g., values of 5 and 10) can be used as triggers for additional management actions. These administrative action levels serve as a basis for instituting additional administrative controls to be sure that doses do not reach the regulatory limit. The protective actions and risk-management approaches taken are situation specific.

Dose proportions using natural background radiation levels as the reference should be interpreted carefully. The natural background radiation level is not constant. It varies with geography. Dose proportions derived from the same occupational dose in one location may be different from that in another location because background radiation levels differ, although variations are likely to be minor. Use of an average natural background may be a reasonable compromise, but this may lead to overestimation or underestimation in specific circumstances. Uncertainties in dose proportion may also lead to interpretive difficulties. When dose limits are used as reference, the only source of uncertainty is the dose estimate since regulatory limits by definition have no uncertainty. Uncertainties in background must be considered when it is used as reference. Background level uncertainties will usually be much smaller than the uncertainties in the measured or calculated doses.[27] When reporting dose proportions, a range of values should be given to reflect uncertainties in the calculations.

Since a linear relation between dose proportion and health effects is not assumed, it would be incorrect to conclude that a dose proportion of 50 is associated with twice the cancer risk as a dose proportion of 25. Dose proportions require that risks be thought of in semiquantitative terms. Higher dose proportions imply greater cancer risks, but the magnitude of the increase is not determinable. This in no way diminishes its utility in radiological protection. The magnitude of cancer risks at low doses is not very well known. At occupational and environmental doses, a tenfold reduction in risk cannot be measured reliably because of large uncertainties in risk estimates. For example, lung cancer risks due to exposure to radon concentrations in air of 400 Bq/m^3 and 40 Bq/m^3 cannot be distinguished.[28]

It is prudent to assume that if risks exist at low doses, increasing dose proportions are likely to be associated with elevated risk. Nevertheless, these risks are still too small to be measured in most occupational settings. Since risk cannot be measured well, the shape of the dose response curve at low doses is inconsequential. Comparisons can be made with regulatory limits to put the dose in a safety context (limits are set at levels well below measurable risk), and comparisons can be made with natural background levels to put the magnitude of the dose in perspective.

The dose proportion is a useful tool to put dose in context.[29] Expressing dose in terms of multiples (or submultiples) of natural background is an effective communication strategy provided that there is clear understanding of what natural background means. Natural background is a complex mixture of different types of radiation from various sources. Only those components of natural background that are comparable to the exposures in question should be used. Dose proportions involving external radiation exposures should use natural background levels, excluding radon because it is an internal emitter.

The proposed dose-based system of protection does not preclude use of conventional risk benefit or cost benefit analyses in decision making. In the context of dose proportion the analyses are simply based on dose rather than risk. In many ways, as discussed previously, this is a preferable approach because dose can be measured to levels well below corresponding risks. Accordingly, uncertainties in dose are smaller and confidence in decisions is higher.

SIMPLIFICATION OF RADIATION QUANTITIES AND UNITS

A dose-based system requires a simple and straightforward set of dose quantities and units. Radiological protection is burdened with a highly complex system of quantities and units. Clearly a single quantity cannot be used in all situations. Average tissue dose is a useful quantity for external exposure to penetrating gamma radiation but has limited utility for internally deposited radionuclides that are distributed nonuniformly. A simplified system of quantities and units can minimize miscommunication and misinterpretation of information. Communicating dose in terms familiar to the public and decision makers is critical to an efficient dose-based system of protection.

The current framework for radiation protection is risk based because doses are determined from risk-derived tissue-weighting factors that allow tissue specific doses to be summed and compared with limits. Tissue-weighting factors are dimensionless and are defined as the proportion of the whole body risk of radiogenic cancer that is attributable to a specific tissue or organ. The product of the measured dose and the tissue-weighting factor is the effective dose.

Use of effective dose as a quantity in radiological protection has become increasingly questioned. For example, determining the whole body equivalent dose from ^{131}I or ^{222}Rn for the purposes of adding effective doses together from different radionuclides is biologically unsupportable because health effects are only observed in target tissues. These radionuclides (and others) need to be treated separately from gamma-emitting radionuclides or radionuclides that emit radiation uniformly throughout the body (or any situation in which the radiation exposure is legitimately a whole-body exposure).

Internal emitters can have complex behaviors in the body. Radon is an example of an agent for which conventional dose measurement is inappropriate. Lung is the target organ for radon and dose is highly nonuniform, making it difficult to interpret dose information. In this case it is more appropriate to use radon concentration in air (Bq/m^3) as a surrogate for lung dose. Activity in air is the quantity that is actually measured.

There are numerous dose-related quantities (categorized as either dosimetric or protection quantities) in use in radiation protection. Debate continues about the stability of radiation protection quantities and units and the appropriateness of protection quantities like equivalent dose. Recent changes in the names of certain quantities have generated confusion among experts and the public. The change from effective dose equivalent to effective dose, and the shift from dose equivalent to equivalent dose have created chaos in the nomenclature.[30] Further, the same units are used for multiple quantities. The sievert is a unit common to both equivalent dose and effective dose. Unless the specific quantity is identified, use of the sievert is problematic. The failure of the U.S. to adopt the modern metric system only adds to the confusion.

The radiation protection community should give serious thought to simplifying the system of radiation quantities and units. Protection quantities such as "equivalent dose" and "effective dose" are not measured directly and are not independent from absorbed dose. Since radiation and tissue-weighting factors are dimensionless, there is no need to use other quantities to describe weighted absorbed dose. Radiation protection should be based on direct measurements of absorbed dose where possible, since it is the energy absorbed per unit mass of tissue that is most closely related to the probability of health effects. An obvious exception is radon because of the nonuniform deposition of radon progeny in the lung.

For protection purposes, the ICRP prefers to use the average tissue dose. The ICRP argues that the absorbed dose (calculated at a point in tissue) is an unsatisfactory predictor of risk because it does not account for the average tissue dose or differences in tissue radiosensitivity. The magnitude of the average tissue dose depends on weighting factors and assumptions in dose models.[31] These factors introduce large uncertainties in the calculations. Since radiation risk estimates have much greater uncertainties at low doses delivered at low dose rates, it is unclear that reducing the uncertainty in the dose estimate by using average tissue dose will have any practical consequence. Accordingly, it is prudent to adopt the simpler approach by using absorbed dose or other appropriate quantity.

The refusal of the U.S. to adopt the modern metric system continues to cause communication problems in the international arena. The U.S. is the only industrialized country in the world not officially using the modern metric system. Because of its many advantages, including easy conversion between units of the same quantity, the modern metric system has become the internationally accepted system of measurement units. The U.S. government should adopt it with the gray (Gy) as the fundamental unit of absorbed dose. The U.S. NRC has already initiated an effort to use the modern metric system. Since 1993, the agency has published new regulations, regulatory guides, and other agency documents in dual units (English system and modern metric system) to facilitate use of modern metrics by licensees.[32] This is not a perfect solution, but it is a worthwhile effort to get licensees and others to use the modern metric system. By doing so, the U.S. will align itself with the rest of the world. We should think globally rather than locally about radiological health and safety.

REVIEW OF THE CURRENT SYSTEM OF RADIATION PROTECTION

The ICRP is reviewing its system of radiation protection and developing new recommendations that will replace the 1990 recommendations.[33] The ICRP Main Commission is now considering what it views as a simpler approach to radiation protection based on an individual-oriented philosophy. The principal change involves emphasis on the dose to an individual from a controllable source. This represents a shift from the utilitarian philosophy emphasizing societal-oriented criteria that are the basis of the current framework. However, it is unclear that abandoning a utilitarian perspective simplifies radiation protection. The proposed radiation protection framework is still unnecessarily complicated. The dosimetric and protective quantities introduced in ICRP's 1990 recommendations are slated for retention, but the next recommendations are expected to clarify differences in quantities.[34]

The ICRP admits that the current set of radiation and tissue-weighting factors is more complex than can be justified. The next set of recommendations will attempt to simplify the weighting factors.[35] The proposed system also introduces a complex generalized structure of individual doses linked to protective actions. The various protective actions are linked to levels of concern (called "Bands") that are defined in terms of multiples and submultiples of the natural background radiation dose. Serious consideration should be given to simplifying quantities used in radiation protection. The proposed use of multiples of the natural background as a basis for protective actions is a sound basis for developing a system of protection based on natural background radiation levels (Table 7.2).

It is time for radiological protection to scrap risk and reinvent dose. The idea of a dosed-based system of protection is not new. The ICRP described such a system in its 1966 recommendations and introduced a risk-based system of protection in its Publication 26 in 1977.[36] The risk-based system was designed to address practical concerns and limitations of the dose-based system, including the inability to sum doses from various radiological sources. However, the risk-based system has turned out to have significant problems as discussed above. Given that no system is perfect, radiation protection should be based on a framework that is scientifically defensible, is conceptually straightforward, and facilitates communication and decision making. The present proposal is conceptually consistent with the older dose-based system. The proposed framework recognizes recent advances in radiobiology and epidemiology, suggesting that use of LNT theory and effective dose are flawed concepts at low doses. Use of dose proportion avoids the need to use LNT or any other theory to predict risk. The recommendation to use absorbed dose, activity concentration, or other suitable metric in determining dose proportion avoids the effective dose problems. Dose additivity remains an unavoidable concern. This, however, is a reflection of the fact that internal deposition of radionuclides and external radiation exposure are fundamentally different. For radiation protection purposes, differences in internal and external radiation exposure cannot be easily resolved and should be considered separately.

NOTES AND REFERENCES

1. American Cancer Society, *Cancer Facts and Figures 2005* , American Cancer Society, Inc., Atlanta, 2005. Department of Health and Human Services, *The Health Consequences of Involuntary Exposure to Tobacco Smoke*: *A Report of the Surgeon General*, Department of Health and Human Services, Centers for Disease Control and Prevention, Coordinating Center for Health Promotion, National Center for Chronic Desease Prevention and Health Promotion, Office of Smoking and Health, Atlanta, 2006.
2. ICRP introduced the new framework for protection in Publication 26 published in 1977. The system was revised and updated in ICRP's current recommendations found in Publication 60. International Commission on Radiological Protection (ICRP), Recommendations of the International Commission on Radiological Protection, Publication 26, Pergamon Press, Oxford, 1977; International Commission on Radiological Protection (ICRP) 1990 Recommendations of the International Commission on Radiological Protection, Publication 60, Pergamon Press, Oxford, 1991.
3. Abelson, P., Uncertainties about health effects of radon, *Science,* 250, (4979), 353, 1990.
4. Risks associated with asbestos exposure cannot be compared or combined with risks from benzene because asbestos causes mesothelioma of the lung and benzene causes leukemia. See American Cancer Society, *Cancer Facts & Figures, 2005* , American Cancer Society, Inc., Atlanta, GA, 2005.
5. *Supra* note 2.
6. International Commission on Radiological Protection (ICRP), Recommendations of the International Commission on Radiological Protection Publication 26, Pergamon Press, Oxford, 1977.
7. U.S. NRC, Standards for Protection against Radiation, 10CFR§20, 1997.
8. United Nations Scientific Committee on the Effects of Atomic Radiation (UNSCEAR), *Sources and Effects of Ionizing Radiation,* UNSCEAR 2000 Report to the General Assembly, with scientific annexes, Volume II: Effects, United Nations, New York, 2000.
9. National Research Council, *Health Effects of Exposure to Low Levels of Ionizing Radiation,* BEIR V Report, National Academy Press, Washington, DC, 1990.
10. National Research Council, *Health Effects of Exposure to Radon* ., BEIR VI Report, National Academy Ruve, Washington, DC, 1992.

11. National Council on Radiation Protection and Measurements (NCRP), *Ionizing Radiation Exposure of the Population of the United States,* NCRP Report No. 93, National Council on Radiation Protection and Measurements, Bethesda, MD, 1987.
12. *Supra* note 10.
13. Brenner, D.J. et al., Cancer risks attributable to low doses of ionizing radiation: Assessing what we really know, *Proceedings of the National Academy of Sciences,* 100, 13761, 2003.
14. In some situations, doses may be calculated because direct measurement is not possible as in the case of radiation doses to target organs and tissues from internally deposited radionuclides.
15. A detailed history of radiation dose limits in the U.S. is provided in Jones, C.G., A review of the history of U.S. radiation protection regulations, recommendations and standards, *Health Physics,* 88, 697, 2005. For comparison, BEIR risk estimates plotted in Figure 7.1 are based on relative risk projection models and assume the LNT dose-response function.
16. *Supra* note 2. In 1990 ICRP recommended that worker limits be set at 100 mSv over any five-year period. This is equivalent to an average of 20 mSv per year with a limit of no more than 50 mSv in any one year.
17. Under the unity rule, regulatory compliance would be satisfied if: $(D_1/L_1) + (D_2/L_2) + (D_3/L_3) + \ldots (D_n/L_n) < 1$ where D/L is the dose proportion (ratio of the measured or calculated dose to the appropriate dose limit) for agents 1, 2, 3…n.
18. Mossman K.L. et al., Bridging Radiation Policy and Science: Final Report of an International Conference, Bridging Radiation Policy and Science Conference, McLean, VA, 2000.
19. Ghiassi-Nejad, M. et al., Very high background radiation areas of Ramsar, Iran: Preliminary biological studies, *Health Physics,* 82, 87, 2002.
20. National Research Council, *Health Risks from Exposure to Low Levels of Ionizing Radiation,* BEIR VII Report, National Academies Press, Washington, DC, 2005.
21. Gribble, G.W., Amazing organohalogens, *American Scientist,* 92, 342, 2004.
22. Adler, H.I. and Weinberg, A.M., An approach to setting radiation standards, *Health Physics,* 34, 719, 1978.
23. Mossman, K.L., The linear no-threshold debate: Where do we go from here?, *Medical Physics*, 25, 279, 1998; Clarke, R.H., Control of low level radiation exposure: Time for a change?, *Journal of Radiological Protection*, 19, 107, 1999.
24. Health Physics Society Position Statement, *Ionizing Radiation-Safety Standards for the General Public,* June 2003. http://hps.org/documents/publicdose03.pdf (accessed March 2006).
25. *Supra* note 8. U.S. federal dose limits are used for the purposes of discussion. Worker exposures are limited to 50 mSv per year; public exposures are limited to 1 mSv per year.
26. Alexskhin, R.M. et al., *Large Radiation Accidents: Consequences and Protective Countremeasures,* IzdAT Publisher, Moscow, 2004.
27. For a discussion of external and internal dosimetry and measurement uncertainties see Poston, Sr., J.W., External dosimetry and personnel monitoring, *Health Physics,* 88, 557, 2005; Potter, C.A., Internal dosimetry: A review, *Health Physics*, 88, 565, 2005.
28. Lubin, J.H. and J.D. Boice Jr., Lung cancer risk from residential radon: Meta-analysis of eight epidemiological studies, *Journal of the National Cancer Institute,* 89, 49, 1997.
29. The 1997 Presidential/Congressional Commission on Risk Assessment and Risk Management also advocates the development of a common metric to assist comparative risk assessment and risk communication. As discussed in this report, the EPA uses a margin-of-exposure (MOE) approach to compare measured or calculated

exposures to an agent against a reference based on exposures that produce a specific health effect. This is similar in concept to the dose proportion discussed in this book. Interpretations of MOEs are made difficult by the uncertainties in the reference exposures. See Presidential/Congressional Commission on Risk Assessment and Risk Management, *Risk Assessment and Risk Management in Regulatory Decision-Making,* Final Report Volumes 1 and 2, 1997. http://www.riskworld.com/riskcommission/Default.html (accessed March 2006).

30. International Commission on Radiological Protection (ICRP), 1990 Recommendations of the International Commission on Radiological Protection, Publication 60, Pergamon Press, Oxford, 1991; Strom, D.J. and Watson, C.R., On being understood: Clarity and jargon in radiation protection, *Health Physics,* 82, 373, 2002.
31. International Commission on Radiological Protection (ICRP), A report on progress towards new recommendations: A communication from the International Commission on Radiological Protection, *Journal of Radiological Protection,* 21, 113, 2001.
32. U.S. NRC, Conversion to the metric system; policy statement, *Federal Register,* 61, 31169, 1996.
33. International Commission on Radiological Protection (ICRP), A report on progress towards new recommendations: A communication from the International Commission on Radiological Protection, *Journal of Radiological Protection,* 21, 113, 2001.
34. International Commission on Radiological Protection (ICRP), 1990 Recommendations of the International Commission on Radiological Protection. Publication 60, Pergamon Press, Oxford, 1991.
35. *Supra* note 33.
36. International Commission on Radiological Protection, *Recommendations of the International Commission on Radiological Protection* . ICRP Publication 9, Pergamon Press Ltd., Oxford, 1966, *Supra* note 6.

8 Radiation from the Gods

The U.S. Environmental Protection Agency (EPA) considers indoor radon to be one of the most important causes of cancer in the U.S.[1] The agency estimates that about 20,000 lung cancer deaths annually are caused by radon exposure in homes.[2] Only cigarette smoking (responsible for about 150,000 lung cancer deaths annually) causes more lung cancer deaths. If the EPA is right, the risk of radon exposure could far exceed hazards associated with typical pollutants in outdoor air, drinking water, and certain foods and would approximate hazards of some more common activities such as automobile travel.[3] Because the entire U.S. population is exposed, radon gas would account for more deaths than almost any other agent the EPA regulates.

Lung cancer and domestic radon exposure provide a clear example of many of the themes and issues discussed in this book. Federal and state governments have aggressively promoted indoor radon as a serious public health hazard because radon is a major contributor to radiation dose, and we have the technological capabilities to do something about high radon levels in houses and buildings. Much of the debate on the public health consequences of radiation exposure centers on the question of uncertainty in risks at small radon levels. Because of the limited data that shows significant risks from epidemiological studies of indoor radon, linear no-threshold theory (LNT) is used to predict risks based on observations made in the occupational (mining) environment where radon levels are thousands of times higher. Communicating radon risks has proven to be a challenging exercise because of the significant public apathy toward radon remediation. Using comparisons of lung cancer deaths from radon and cigarette smoking has proven to be a less than satisfactory communication strategy, in part because smoking and radon exposure are unrelated activities. Public health impacts of radon exposure in homes are not clearly established because risks are highly uncertain. For more than 95% of U.S. homes, risks are either too small to be measured reliably or are essentially zero. The residential radon problem is a clear example of the need to adopt a dose-based system of protection whereby radon levels in homes are compared to natural background levels as a basis for public policy and decision making. A dose-based system avoids the need to quantify risks that are highly uncertain. Except for homes that have very high radon levels, decisions to remediate based on risk are highly questionable. Finally, managing radon risks and costs of home remediation highlight the important problem of allocating limited economic resources to manage a public health problem (lung cancer) that has other more significant roots.

Radon is the single largest source of human exposure to ionizing radiation (Table 8.1). It accounts for about half of the total average annual dose from natural background radiation and represents about 40% of the exposure from all sources including medical uses.[4]

TABLE 8.1
Sources of Radiation Exposure

Source	Average Annual Effective Dose (mSv)
Cosmic rays	0.4 (range: 0.3–1.0)
External terrestrial radiation	0.5 (range: 0.3–0.6)
Inhalation (mainly radon gas)	1.2 (range: 0.2–10.0)
Ingestion	0.3 (range: 0.2–0.8)
Medical diagnostic	0.5 (range: 0.1–25.0)

Sources: National Council on Radiation Protection and Measurements, *Exposure of the U.S. Population from Diagnostic Medical Radiation*, NCRP Report No. 100, NCRP, Bethesda, MD, 1989; United Nations Scientific Committee on the Effects of Atomic Radiation, *UNSCEAR 2000 Report to the General Assembly, with Scientific Annexes, Volume I: Sources*, United Nations, New York, 2000.

Doses from medical applications of radiation and from radon have particularly wide ranges. Medical doses depend on the size of the patient and the nature of the diagnostic study. Simple chest x-rays result in doses approximating 0.5 mSv; computerized tomography (CT) studies (particularly in children) and fluoroscopic examinations may deliver doses of several tens of mSv. Factors contributing to variations in radon levels are related to local geological and housing characteristics.

The EPA's estimate of 20,000 lung cancer deaths annually from domestic exposure to radon comes from extrapolating risks derived from statistically significant data from occupational exposures in uranium and other metal mining environments. Miner studies have clearly established radon gas as a cause of lung cancer, although uncoupling radiogenic risk from smoking risk has been problematic since many miners smoked. The interaction of risks from cigarette smoking and radon exposure appears to be more than additive, but the exact nature of the interaction remains unclear. Demonstrating that lung cancer risk is elevated due to domestic radon has proven to be enormously difficult because radon levels are thousands of time lower than in the mining environment. Communicating radon risks has also been challenging because of public apathy.

Radon as a domestic health issue is a relatively new concern. For several decades radon gas has been a known cause of lung cancer. However, the hazard was believed to be restricted to underground mining settings where very high concentrations could accumulate. Radon gas is measurable in homes, but governments and technical experts never realized that high concentrations similar to that found in mines could accumulate in homes. The saga of Stanley Watras and his home in Pennsylvania changed all that.

THE WATRAS CASE

Stanley Watras lived near Harrisburg, PA, and worked as an engineer at the nearby Limerick Nuclear Power Plant.[5] When exit radiation portal monitors were installed at the plant, Watras continuously tripped the exit portal radiation monitor on his way to off-site meetings and at the end of his shift. What made this troublesome was the fact that Limerick was not yet operational. Furthermore, analysis of the radioactive

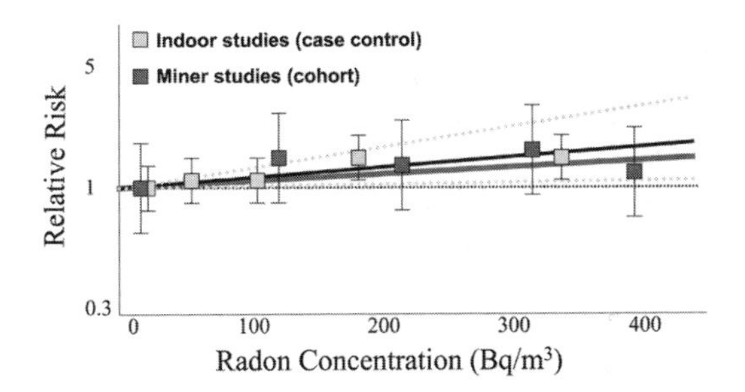

FIGURE 8.2 Lung cancer risks and radon. Mortality risk from lung cancer is a linear function of radon concentration. Relative risks (RR) are derived from the meta-analysis of indoor radon studies and pooled analysis of underground miner studies. The light line (and accompanying light data points with 95% confidence limits) is the fitted exposure response from indoor radon studies; the dark line (and accompanying dark data points with 95% confidence limits) is the fitted exposure response from miner studies. The line at RR = 1 indicates no health effect from radon exposure. (Modified from Lubin, J.H. and J.D. Boice, Jr., Lung cancer risk from residential radon: Meta-analysis of eight epidemiological studies, *Journal of the National Cancer Institute,* 89, 49, 1997.)

of zero risk. When combined effects between radon and smoking are considered, almost all of the radon risk accrues to smokers. Accordingly, it is unclear whether the domestic environment poses a significant health risk.

The large background rate of lung cancer mortality due to cigarette smoking dwarfs any radiogenic risks. The EPA's population estimate of about 20,000 lung cancer deaths per year due to radon contains substantial uncertainty because of the confounding effects of cigarette smoking and because the estimate is the product of mapping a highly uncertain radiogenic risk onto the entire U.S. population.[12]

Figure 8.2 shows data from a pooled analysis of 11 miner studies and meta-analysis of residential radon studies. An LNT dose response can be fit to the miner and residential data with a positive risk coefficient.[13] However, it is also reasonable to conclude that the slope of the linear dose response is consistent with zero and that radon has no effect on lung cancer risk for radon concentrations up to about 400 Bq/m^3 because statistical uncertainties are large. Fewer than 2% of U.S. homes have radon concentrations greater than 400 Bq/m^3. The large statistical uncertainties in the indoor radon data do not preclude zero health risk for radon concentrations less than 400 Bq/m^3.

In spite of the somewhat consistent findings of miner and residential radon studies, questions have been raised about whether radon is actually harmful at low levels. Bernard Cohen, a physicist at the University of Pittsburgh, argues that low levels of radon do not increase lung cancer deaths and instead appear to be beneficial by lowering death rates. Hormesis proponents used Cohen's findings to support their position that the LNT theory is wrong and that low-level radiation is beneficial to health. Cohen's approach differed significantly from more traditional epidemiological methods by using an ecological method to correlate residential radon levels and lung cancer mortality in 1600 U.S. counties.[14] This represents the single largest source of data on health effects of radon.

Cohen's data clearly predict outcomes paradoxical to miner and residential studies. Instead of a positive slope (as shown in Figure 8.2), Cohen's data are consistent with a biphasic dose-response function characterized by an initial negative slope (low concentrations of radon reduce the risk of lung cancer) followed by a positive slope at higher concentrations of radon (high concentrations of radon increase the risk of lung cancer). Subsequent analyses of Cohen's data showed that cigarette smoking was not completely accounted for and the initial negative slope of the dose response could be explained by the overwhelming confounding of cigarette smoking.[15]

The National Research Council's (NRC) BEIR VI committee's estimate of 15,400 lung cancer deaths per year attributable to radon is the committee's best assessment based on currently available data from miner studies. Uncertainty analysis indicates that the number of cases could range from as few as 3,000 to as many as 32,000 (95% confidence limits) lung cancer deaths per year.[16] However, because of the large uncertainties the lower limit of uncertainty should include zero.[17]

The BEIR VI report conclusions are entirely consistent with the current EPA radon policy by suggesting that: (1) radon exposure in the domestic environment is a substantial public health hazard, and (2) reduction of indoor radon concentrations to levels at or below the EPA action guide (150 Bq/m^3) may avoid a significant number of lung cancer deaths. However, epidemiological evidence to support these conclusions is either absent or not convincing. A more reasonable conclusion is that lung cancer risk is insignificant for radon concentrations below 400 Bq/m^3 (Figure 8.2).

The EPA's own estimates are based on the BEIR VI report. According to the EPA, radon accounts for about 13% of all lung cancer mortality. About 26% of all lung cancers in never smokers is radon related. EPA further estimates that 26% of the total risk would be avoided if all homes above 150 Bq/m^3 were reduced to 40 Bq/m^3.[18]

In spite of the fact that indoor radon is now recognized as the most hazardous pollutant in the indoor environment, there is a great deal of uncertainty underlying the severity of the so-called "radon problem." There are no epidemiologic studies that provide direct, unequivocal evidence of increased lung cancer mortality at environmental levels of radon. The health hazard of indoor radon is primarily inferred from epidemiologic studies of miners exposed to high radon levels.

IS THERE REALLY A PUBLIC HEALTH HAZARD?

Public Health

Twenty thousand lung cancer deaths per year is a very large number and suggest that radon is a serious public health hazard. But is it? How realistic is the EPA's estimate of lung cancer mortality? The EPA uses LNT theory to predict the number of cancer deaths based on the size of the population and risks derived from uranium miner studies. Multiplying a very large population by a small individual risk yields a large number of cancer deaths, even though only a very small percentage of people are subject to high radon exposures (Figure 8.1). Should residential radon be considered a public health hazard because the entire U.S. population is exposed to radon, even though average individual risks are very small? At average residential levels (approximately 50 Bq/m^3), radon is associated with a lifetime risk of lung cancer death of about 0.2%.[19] This theoretical risk is small compared to health risks of smoking and obesity but is larger than the risks of many agents regulated by the EPA, including benzene, chloroform, and ethylene dibromide.

The EPA's radon responsibilities are codified in the Indoor Radon Abatement Act of 1988. The legislation establishes a long-term goal that indoor air be as free from radon as the ambient air outside buildings. This is a laudable goal but is technically not feasible — home construction cannot entirely eliminate suboptimal ventilation rates and differences in indoor and outdoor air pressure. The law authorizes funding for radon-related activities at the state and federal levels.[20] Individual states such as Pennsylvania and New Jersey now have extensive experience working with communities to provide home radon tests and to assist with remedial action programs. The EPA also established standards for radon testing and home remediation to provide some assurance to homeowners that radon companies use government-approved methods.

The EPA has established an indoor air action guideline of 150 Bq/m^3.[21] The EPA's action level was based on guidelines established by the U.S. Department of Energy (DOE) to remediate high radon homes located on uranium mill tailings in Grand Junction, Colorado. The level was also established based on considerations of what is technologically feasible in home radon remediation. Above the action level, EPA recommends that the homeowner consider possible remedial actions to reduce radon levels. The urgency in doing this depends on the radon levels found in the home. Unfortunately, the EPA guide is interpreted incorrectly as law or as a bright line indicating safety. The EPA guideline is just that — a guideline. The EPA recommends remediation if radon concentrations are above the recommended level, but homeowners are not required to do anything except in localities where real estate laws require home sellers to test their home and remediate as a necessary part of a real estate transaction. Homeowners have also assumed that the EPA has determined that homes with radon levels above 150 Bq/m^3 are unsafe. The guideline in no way implies safety or an unhealthy atmosphere. It is strictly an action threshold whereby renovations of the home may be advisable to reduce the radon concentrations further. There are no epidemiologic studies indicating that the 150 Bq/m^3 action level is detrimental to the public health.

Comparing radon levels in homes to natural background levels is a useful metric for remediation decision making. A dose-based system avoids the need to estimate highly uncertain risks.[22] Guidelines for remediation are easily expressed as dose proportions where the EPA action guide of 150 Bq/m^3 is equivalent to a dose proportion of about 4, assuming a natural background level of radon of 40 Bq/m^3. According to EPA policy, homes with dose proportions greater than 4 would be candidates for remediation. It should be reemphasized that the EPA action level of 150 Bq/m^3 was based on technical feasibility rather than any consideration of health risks. Remediation is not likely to provide public health benefit for homes with dose proportions less than 10 (Figure 8.2).

A simple decision framework is shown in Table 8.2. Dose proportions are the home radon concentration in air (in Bq/m^3) divided by the U.S. average residential radon level in homes (assumed to be 40 Bq/m^3).

TABLE 8.2
Remediation Based on Dose Proportions

Dose Proportion	Likely Effects	Action
> 20	Observable risk of lung cancer	Remediation recommended
10–20	Low risk of cancer	Consider remediation
< 10	Risk is either zero or too small to be measured reliably	No action necessary

Dose proportions are somewhat subjective but based on current dose-response data summarized in Figure 8.2. There is little justification for remediation below about 400 Bq/m^3 (a dose proportion of 10). Any remedial action at radon concentrations of 400 Bq/m^3 or lower is not likely to have measurable benefit in terms of reduced lung cancer risks. On the other hand, a dose proportion of 20 or more indicates significant radon concentration that would justify remedial actions. Dose proportions between 10 and 20 represent a gray area and decisions to remediate would be made on a case-by-case basis. Only a few percent of U.S. houses fall in the high-dose proportion category. Regardless of the dose proportion, the decision to remediate is a personal one. Dose proportion information provides important guidance, but the homeowner's perception of the risk and willingness and ability to pay remediation costs are important drivers.

Perceptions and Fears

There is widespread public apathy regarding radon exposure in homes.[23] The public response is perplexing because the public health implications of indoor radon exposure would appear to be serious, according to EPA estimates. Part of the problem is that radon is odorless, colorless, tasteless, invisible, and exposure causes no proximal health effects. It does not cause visible damage to the home — no discoloration of walls or collection in piles like loose asbestos. Public and EPA perceptions of environmental problems differ. In 1992, the EPA considered radon, ozone, and air pollution as the worst environmental problems. The public cited local problems such as hazardous waste sites, exposure to toxic chemicals, and water pollution as the most serious problems.[24]

Perhaps the reason for this general apathy is the fact that radon is a natural problem resulting from the decay of radioactive material in the earth's crust. It is unrelated, for the most part, to technological activities. Other environmental problems associated with hazardous waste sites are directly related to technologic activities. The general public perceives these problems as very serious. If technology created the problem, then technology ought to solve it. With radon there is no one to blame for high radon levels. No specific business or industry can be targeted to seek relief from high radon levels.

Since 1984, radon has received considerable public attention. But the public continues to deny the seriousness of the radon problem in spite of the EPA's national radon campaigns. The EPA has used aggressive tactics to scare the public into taking action. Billboard messages such as "Call 1-800-RADON!," and "Radon is a health hazard in your home," have appeared, as have Ad Council messages in broadcast and print media. In 1988, together with the U.S surgeon general, the EPA issued a nationwide health advisory urging that every dwelling in the country be tested for radon. Testing every home in the country would cost the public billions of dollars and the results of onetime tests would be difficult to interpret because they are highly influenced by time of day, season, and temperature when testing is done. In a nationwide radon campaign many homeowners will be alarmed by unusual high readings because of testing conditions and other factors, and will incur additional, more expensive testing to verify accurate radon levels.

The EPA uses risk comparisons to put residential radon risks in perspective. Risk comparison can be a powerful means of communicating risks to the public if the comparisons are valid. To indicate the seriousness of the radon problem, the EPA compared radon risks with other risks familiar to the public (Table 8.3). Risks are not comparable because they are not measured in the same way. Comparing theory-derived risks of radon exposure with risks of other activities that are based on direct observations is problematic. Deaths due to drunk driving, fires, and falls in homes are directly observable, and the deaths can be clearly linked to the activity. In homes, cancer cannot be easily linked to radon exposure. The EPA has also used cigarette smoking as a basis for comparison. Smoking is also a problematic comparison because equating numbers of cigarettes and radon concentrations in air are unrelated activities and not very meaningful to the public.

A more effective communication strategy uses dose proportions as a metric for comparing radon doses (i.e., concentrations of radon in air) from different sources or with standards or guidelines. Home radon readings can be compared to average readings for the region, state, or nation. Average indoor radon levels are known for almost every county in the U.S. Comparisons may also be made with the EPA action level or with radon levels that are known to be associated with significant health risks. Thus a reading of 40 Bq/m^3 is about equal to the national average, about four

TABLE 8.3
Risk Comparisons

Activity	Deaths/Year
Exposure to residential radon	15,000–22,000
Drunk driving	12,200
Fall in the home	9,300
Fire	3,200
Drowning	900

Sources: U.S. Environmental Protection Agency, *A Citizen's Guide to Radon*, 2nd ed., EPA Air and Radiation Report ANR-464, U.S. EPA, Washington, DC, 1992; U.S. Environmental Protection Agency, *EPA Assessment of Risks from Radon in Homes,* EPA 402-R-03-003, U.S. EPA, Office of Radiation and Indoor Air, Washington, DC, June 2003.

times lower than the EPA action level of 150 Bq/m^3, and at least ten times lower than levels associated with measurable health risks. These are meaningful comparisons that put a particular house reading into perspective for the homeowner without having to calculate uncertain risks.

Economic Impacts

According to EPA guidance, homes with radon levels in excess of 150 Bq/m^3 should be remediated. This requires that every home be tested for radon because that is the only way to identify which ones have high radon levels. Home testing is simple and inexpensive, but results must be interpreted with caution. Radon levels depend on a number of factors including where the detectors are placed, the time of year testing is conducted, weather conditions at the time of testing, and length of the testing period. Basement readings are higher than readings on upper floors; readings tend to be higher in winter than in summer; long-term readings are more reliable than short duration tests. If initial testing results are above the EPA action level, follow-up testing should be performed to verify earlier results.

Surveys of single-family homes indicate that homes can have very high indoor radon levels. Based on the known distribution of radon levels in homes (Figure 8.1), 3 million to 4 million U.S. homes would be expected to have radon levels in excess of the EPA action guide assuming a U.S. housing stock of 70 million homes. Of these, perhaps 1 million homes have high enough readings (in excess of 400 Bq/m^3) to justify remediation. Communities with high radon levels have been identified along geological settings with high soil uranium levels such as the Reading Prong in Pennsylvania. The location of a particular house in an area with high radon levels does not necessarily predict that the home will have high radon. In fact, high radon levels have been measured in some homes located in areas with low uranium deposits. However, in general, homes are more likely to have high radon levels if they are built in areas with soils characterized by high uranium content.

Home radon testing is clearly a personal decision. As demonstrated in the Watras case, high radon levels in one home do not necessarily predict levels in neighboring homes. Although the soil content of uranium and radium is a key determinant of radon levels in the home, it is by no means the only factor governing radon levels. Other factors that enhance radon levels in homes include: the presence of basements rather than crawlspaces, houses with walls below grade, houses with exposed earth in the basement area, houses exposed to the wind, houses built on slopes or ridges rather than in valleys, houses with forced air heating systems, tightly sealed rather than drafty houses, and houses with water supplied from private wells. If a home has one or more of these characteristics, then there may be enhanced radon levels. However, the only way to tell for sure is to make indoor radon measurements.

The EPA argues that there is no known safe level of radon and therefore some risk always exists. Thus, achieving even a small reduction in annual lung cancer mortality through remediation would be worthwhile because lung cancer is an important disease particularly in developed countries. Risks can be reduced by lowering radon levels in the home through appropriate remediation methods. In some instances, sealing cracks in floors and walls may be all that is necessary. In other cases more extensive sub-slab depressurization methods including use of pipes and fans may be required. The appropriate radon reduction method depends on the design of the home and other factors. Remediation costs depend on the severity of the problem and the radon reduction methods employed. On average the cost is roughly \$1,000 per home.[25] For some problem homes, extensive renovation costs could push the tab to \$25,000.

Remediation of homes with radon levels below 400 Bq/m³ will not have an appreciable health benefit for the occupants. There is little risk reduction at these radon levels (Figure 8.2). But remediation costs may be substantial. For homes with very high radon levels, remediation may reduce risks significantly at a small cost. However, remediation costs increase dramatically when radon levels are small to begin with (Figure 8.3). As radon levels (and associated health risks) diminish, costs increase to achieve further reductions.

The impact of the EPA action guidelines on the real estate and housing industry may be substantial. In some states, homeowners are required to have radon tests as a condition of sale. Homeowners may have to pay for remediation work to reduce radon levels below the EPA action level before the sale can be completed. In some

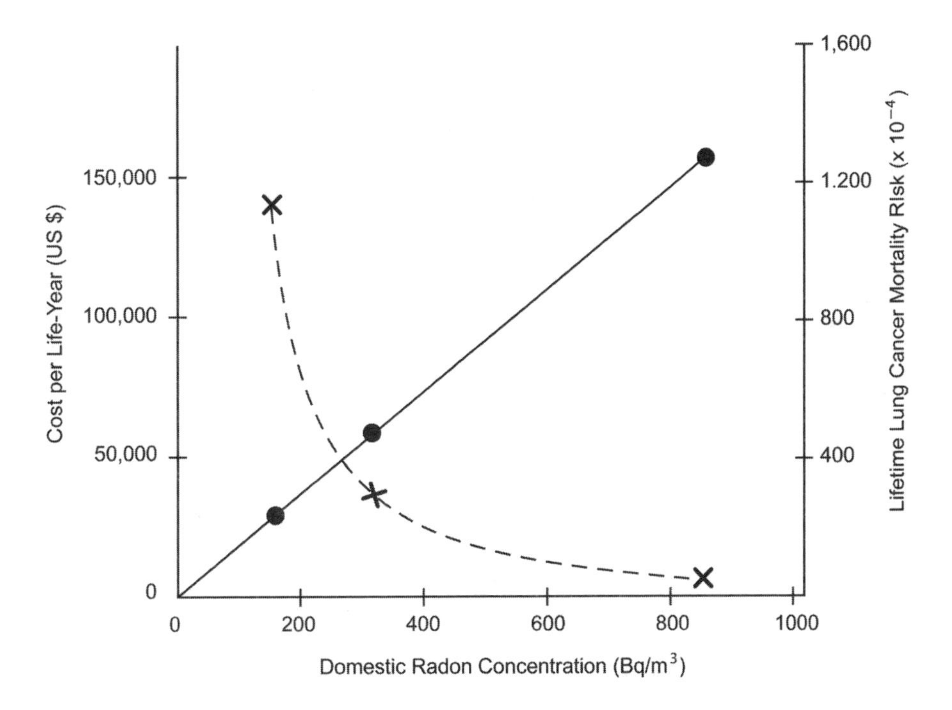

FIGURE 8.3 Economic costs of risk reduction. Costs of reducing environmental and public health effects of ionizing radiation exposure can be enormous. Below a radon concentration of 300 Bq/m³ (about 98% of U.S. homes have radon levels below this level), the costs of further averting radon-induced lung cancer mortality (dashed line) rise significantly despite very small theoretical reductions in lifetime radiogenic lung cancer risk (solid line). (From Mossman, K.L., Restructuring nuclear regulations, *Environmental Health Perspectives,* 111, 13, 2003.)

instances prospective buyers have voided sales because of high radon levels and perceived health risks. New home construction may also be subject to expensive guidelines to reduce radon levels. Today many homes are built to prevent radon from coming in. Radon-resistant construction features usually keep radon levels below about 80 Bq/m^3.

The most effective way to reduce lung cancer deaths attributable to radon is to reduce smoking. The NRC's BEIR VI Committee estimates that about 90% of radon-related lung cancers occur in smokers.[26] The U.S. government does not regulate adult cigarette smoking, although many communities have passed smoking bans in public buildings. The government requires that tobacco products exhibit warning labels concerning health risks of smoking. Radon reduction in new construction and remedial action programs in existing housing may involve expenditures of about $3 billion to reduce the indoor radon concentration with questionable impact on lung cancer incidence. This money would be more wisely spent if it were directed at anticigarette smoking campaigns, the results of which would drastically reduce lung cancer mortality. Although this recommendation is reasonably sound from a scientific and public health standpoint, widespread implementation may not be politically possible.

National/Regional Differences

National radon guidelines and recommendations vary considerably. Governments use the same epidemiological data to estimate risks and rely on the same reports from authoritative bodies. The International Commission on Radiological Protection (ICRP) and the United Nations Scientific Committee on the Effects of Atomic Radiation (UNSCEAR) provide guidance in developing radiation protection policies and standards.

The organization of a radiological protection program depends on several important political, social, and economic considerations. Depending on the political institutions and structures in place, protection programs could be centralized in a federal program or decentralized in a state, provincial, or regional organization. The commitment to radon control also depends on national concerns about public health, particularly cancer. Since radon is a natural product of the environment, a national commitment to environmental issues would be a driving force for an effective radon control program. How the radon risk is socially contextualized would dictate how seriously the radon problem should be taken. In countries where infectious diseases and starvation are the overwhelming health problems, there is little room for concerns about radon.

The manner in which radon is regulated and the resources available to radiological safety depend on economic, social, and political factors. The U.S., Canada, Western Europe, Japan, and Australia enjoy relatively sophisticated regulatory programs, a reasonable number of competent and well-trained professionals, and economic and technological resources. The aims of the government, its resources, the nation's culture, and its overall priorities determine how resources will be used for the benefit of their society as a whole. Most, if not all, Third World countries have no radon guidance or protection programs. These countries do not have the resources available to establish national or regional radon programs. More importantly, lung cancer from indoor radon is not a concern because of the overwhelming public health problems in the Third World involving water quality, infectious diseases, and starvation. Countries with national radon programs are relatively rich, can afford to worry about radon-induced lung cancer, and are able to do something about it.

The U.S. has perhaps the most restrictive radon guidance. Other countries such as Canada, the United Kingdom (U.K.), and Sweden have action levels ranging between 200 and 800 Bq/m^3. National average radon levels and variations in levels have little to do with the setting of action levels or radon regulations. What appears to be important is whether the established goals are politically acceptable and technically achievable and whether sufficient economic resources are available to administratively oversee a radon program on a national or regional level.

In the U.S. radon programs are decentralized. The primary effort for radon measurement and remediation guidance resides at the state level with a strong federal program to support the efforts of the states. The EPA has tackled the radon problem forcefully because of public apathy to radon. The agency's aggressive tactics may also be a reflection of a history of lack of public trust in the agency. In the early 1980s the EPA suffered a poor public image in part because of mismanagement of environmental cleanup programs such as Superfund. A conservative handling of the radon issue, by setting a low level for action and an aggressive public campaign, was viewed as an important way for the agency to restore public trust.

Other countries such as Canada, the U.K., and Sweden have more centralized radon programs. Like the U.S., these countries have a strong commitment to environmental issues, and their national health programs are generally characterized by aggressive approaches to dealing with smoking and other public health threats. National government health and environmental regulatory programs were well established and were well prepared to deal with the radon issue, including setting up public health programs, remediation guidelines, and radon detection.

SMOKING IS THE PROBLEM

Radon gas is the single largest natural source of ionizing radiation exposure to human populations. Radon gas can accumulate to high concentrations in houses that are poorly ventilated. EPA estimates that radon accounts for about 20,000 lung cancer deaths annually. If the EPA's estimate is correct, radon is near the top of the list of important human carcinogens. The Indoor Radon Abatement Act of 1988 authorized the EPA to lead the national effort on radon control. About 5% of U.S. houses (3 million to 4 million homes) have concentrations exceeding the EPA's action limit of 150 Bq/m^3 (assuming a U.S. housing stock of 70 million homes). This is a substantial number of houses. If every home was to undergo some renovation, it could cost the public more than $3 billion (assuming an average renovation cost of $1,000 per house). Whether houses near the EPA action guide should be renovated is arguable. For the 98% of U.S. homes that are below 400 Bq/m^3 there would appear to be no public health problem. There is little health benefit to be gained by remediating these homes since health risks are already very small and there is little change in risk with decreasing radon levels (Figure 8.2). Clearly homes that far exceed the EPA action level (for example homes with radon levels at 800 Bq/m^3 or higher) have high risks and should be remediated to reduce risks. Methods need to be developed to identify homes with high radon levels without having to measure every home in order to find the "needle in the haystack." Of course, public health efforts should concentrate primarily on reducing cigarette consumption. That is the single most important strategy for reducing the lung cancer burden in the U.S.

NOTES AND REFERENCES

1. Radon is an odorless, tasteless, invisible gas that originates from the radioactive decay of uranium in the earth's crust. For a discussion of sources of radon exposure, see Eichholz, G.G., Human exposure, in Cothern, C.R. and Smith, J.E. (Eds.), *Environmental Radon*, Plenum Press, New York, 1987, 131.
2. U.S. Environmental Protection Agency, *EPA Assessment of Risks from Radon in Homes,* EPA 402-R-03-003, U.S. EPA, Office of Radiation and Indoor Air, Washington, DC, June 2003.
3. Nero, A.V., Controlling indoor air pollution, *Scientific American,* 258, 42, 1988.
4. National Research Council, *Health Risks from Exposure to Low Levels of Ionizing Radiation,* BEIR VII Report, National Academies Press, Washington, DC, 2005.
5. This description of the Watras case was taken in large part from Reilly, M.A., The Pennsylvania Experience with Indoor Radon, a paper presented at the Atomic Industrial Forum Conference on Nuclear Industry Radiation Issues: 1986 and Beyond, October 8, 1986; Gerusky, T.M., The Pennsylvania Radon Study, *Journal of Environmental Health*, 49, 197, 1986.
6. Radon concentration is measured in becquerels (Bq) per cubic meter of air (Bq/m^3). One Bq = 1 disintegration per second of radon and its radioactive progeny.
7. In addition to the composition of the soil, other factors that appear to be important in determining radon levels are: (1) the permeability of the gas through the soil, (2) the structure of the home or building must have openings (cracks, crevices, etc.) to allow radon to enter, and (3) the atmospheric pressure inside the home must be less than the pressure outside the home so that radon can move from outside to the internal environment. The most important of these factors appears to be the atmospheric pressure differential. With low atmospheric pressure, the radon emanation rate into the home greatly increases. It is as if the home is literally sucking the radon from the ground. Pressure differentials are produced by the so-called "stack" effect, which is the tendency of air to rise when it is warmer than the surrounding air. Other factors controlling radon emanation rate into the home are associated with the soil surface. For instance, snow cover or standing water may reduce the emanation rate. The factors may lead to marked diurnal and seasonal fluctuations at a particular geographic location. In general, radon levels in homes are higher in winter than in the summer and higher in the morning than afternoon. However, the total variation is less than a factor of 10.
8. American Cancer Society, *Cancer Facts and Figures 2005,* American Cancer Society, Inc., Atlanta, 2005.
9. For detailed discussions about the health effects of radon and particularly epidemiological studies of uranium miners that form the basis for health risk estimates see National Research Council, *Health Effects of Exposure to Radon,* BEIR VI Committee Report, National Academy Press, Washington, DC, 1999.

10. Ibid. The most recent analysis of radon risk estimates was conducted by the National Research Council in its BEIR VI report. Populations that have been evaluated include the uranium miners on the Colorado Plateau; uranium miners in Ontario, Canada; uranium miners employed at the Eldorado Mine at Beaver Lodge, Saskatchewan, Canada; and iron miners at Malmberget, Sweden. In addition to these groups, studies have also been conducted on Czechoslovakian uranium miners and Newfoundland, Canada, fluorspar miners. One of the largest and most important populations is the Colorado Plateau uranium miners. Over 3,000 workers have been followed for an average of 25 years. A total of 256 lung cancer deaths have been registered in this mining population.
11. Lubin, J. and Boice, J.D., Lung cancer risks from residential radon: Meta-analysis of eight epidemiological studies. *Journal of the National Cancer Institute,* 89, 49, 1997; Krewski, D. et al., Risk of lung cancer in North America associated with residential radon, *Epidemiology,* 16, 137, 2005; Lubin, J. et al., Risk of lung cancer and residential radon in China: pooled results of two studies, *International Journal of Cancer,* 109, 132, 2004; Darby, S. et al., Radon in homes and risk of lung cancer: collaborative analysis of individual data from 13 European case-control studies, *British Medical Journal,* 330(7485), 223, 2005; Darby, S. et al., Residential radon and lung cancer — detailed results of a collaborative analysis of individual data on 7,148 persons with lung cancer from 13 epidemiologic studies in Europe, *Scandinavian Journal of Work Environment and Health,* 32 (Suppl. 1), 1, 2006.
12. *Supra* note 2. The EPA's estimate is based on 1995 cancer mortality statistics and assumes a total U.S. lung cancer mortality burden of 157,400 per year. The corresponding 90% confidence intervals for radon-induced lung cancer in 1995 are 8,000 to 45,000.
13. *Supra* note 9 (BEIR VI report). The National Research Council's BEIR VI Committee used the LNT theory to predict lung cancer risks from radon exposure in homes based on measured risks in uranium miners exposed to radon levels that were on average thousands of times higher.
14. The ecological method is a generalized epidemiological method that involves the study of groups rather than individuals. It is therefore much less powerful than case control or cohort analyses in risk assessment. In the ecological approach it is not possible to determine who in the group has the disease in question and what their exposures were. Furthermore, the contributions of confounding factors such as cigarette smoking cannot be fully evaluated. See Cohen, B., Test of the linear no-threshold theory of radiation carcinogenesis in the low dose, low dose-rate region, *Health Physics,* 68, 157, 1995.
15. Puskin, J., Smoking as a confounder in ecological correlations of cancer mortality rates with average county radon levels, *Health Physics,* 84, 526, 2003.
16. *Supra* note 9.
17. The 1990 BEIR V Committee, in its review of health effects of low-level radiation, concluded that the possibility that there may be no health risks from radiation doses comparable to natural background radiation levels (e.g., levels of radon in typical homes) cannot be ruled out. At low doses and dose rates, the lower limit of the range of statistical uncertainty includes zero. See National Research Council, *Health Effects of Population Exposure to Ionizing Radiation,* BEIR V report, National Academy Press, Washington, DC, 1990.
18. *Supra* note 2.
19. EPA report presents new radon risk estimates, *Health Physics News,* 32(1), 1, January 2004.

20. The Indoor Radon Abatement Act of 1988 authorizes the EPA to administer grants to help states establish radon programs, conduct radon surveys, develop public information on radon, and conduct demonstration and mitigation projects. The Act also requires the EPA to provide technical assistance to states. The EPA would assist states with radon surveys, training seminars, mitigation projects, development of measurement, and mitigation methods for nonresidential child care facilities, and public information materials.
21. U.S. EPA, *A Citizen's Guide to Radon*, second edition, EPA Air and Radiation Report ANR-464, U.S. EPA, Washington, DC, 1992.
22. Radon concentration in air (in units of Bq/m^3) is used as a surrogate for lung dose. Activity concentration in air is a straightforward and reproducible measurement. Assessing radon dose to the lung is problematic in part because radon and its radioactive progeny are distributed nonuniformly in the lung.
23. Cothern, C.R., Widespread apathy and the public's reaction to information concerning the health effects of indoor air radon concentrations, *Cell Biology and Toxicology*, 6, 315, 1990.
24. Farhar, B.C., Trends in U.S. public perception and preferences on energy and environmental policy, *Annual Review of Energy and the Environment*, 19, 211, 1994.
25. *Supra* note 21. Cost estimates are based on 1992 dollars. At an inflation rate of 3% per year costs would be closer to $1,500 per home in 2005 dollars.
26. *Supra* note 9.

9 Hold the Phone

Almost everyone uses a cell phone, and almost no one thinks it is dangerous. Cell phones have transformed business and social activities in ways that few people could have predicted 15 or 20 years ago. The technology is now completely embedded in our social fabric. But as with any new technology, safety concerns have been raised. Although clear scientific evidence of cancer risk is lacking, some governments have issued directives to control cell phone use. Such policies have only served to heighten public fears but not to the point that cell phone use has diminished. The public health impact of cell phones serves as a powerful case study to illustrate themes in this book related to risk assessment and risk management. What constitutes a public health threat? How well can we measure small risks when experiences and direct observations are limited? Should a precautionary approach to risk management be adopted when evidence of risk is lacking? The government of the United Kingdom (U.K.) has issued a directive that children up to the age of about 16 should avoid using cell phones as a precautionary measure. Invoking the precautionary principle in this way has created serious questions about management of essentially phantom risks.

Cell phone use has grown exponentially in the U.S.; since 1985, use of cell phones has grown at the rate of about 15% per year. This means that the number of cell phones doubles approximately every 5 years assuming an exponential growth model.[1] The rate of growth is likely to slow in the near future as population use becomes saturated. In 2005 there were more than 180 million cell phone users in the U.S.

Cell phones are nothing more than two-way radios that use electromagnetic radiation in the radiofrequency (RF) range to transmit and receive signals. What distinguishes RF from visible light and other kinds of electromagnetic radiation is the radiation frequency (Figure 9.1). Cell phones and other personal communications systems operate in the 300 to 3000 megahertz (MHz) range (1 MHz equals 1 million cycles per second).

Cell phone communications are supported through a network of cellular towers. These are the base stations for the cells that make up the coverage area. Towers can be as high as 200 feet tall. The antennas at the top of the towers emit radiofrequency electromagnetic waves but at levels much lower than commercial radio and television stations. The radiofrequency energy these cellular antennas radiate has a power level similar to the energy emitted by common household light bulbs. The strength of these electromagnetic fields decreases rapidly with distance from the antennas. Human exposures typically occur 100 meters or more away from the tower. In some locations, the strength of the radiofrequency fields can decrease to almost undetectable levels at site property lines. Electromagnetic field intensities from cell phone towers are well below regulatory limits. Even at peak intensity, levels are thousands of times lower than the limits set for human exposure to radiofrequency fields.

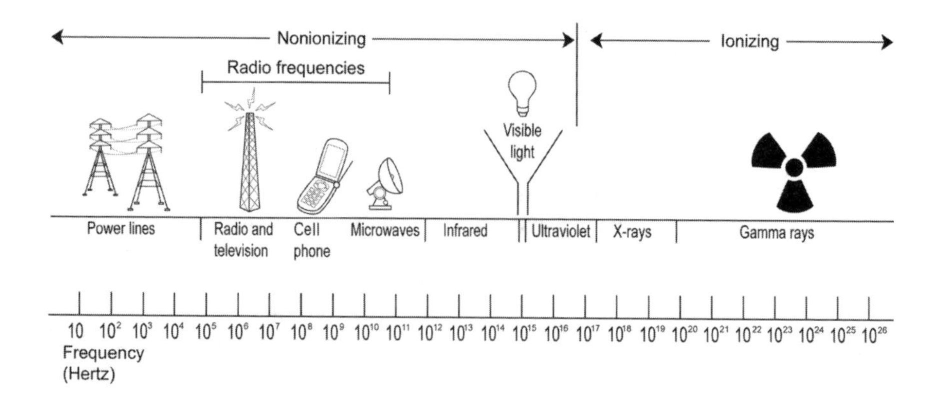

FIGURE 9.1 The electromagnetic spectrum. Cell telephones transmit and receive electromagnetic radiation in the radiofrequency (RF) range. RF radiation is in the nonionizing portion of the electromagnetic spectrum. The radiation energy is low and insufficient to cause ionization. On the other hand, x-rays and gamma rays have sufficiently high energy to cause ionization that may result in serious biological damage. The human eye detects a very small portion of the nonionizing electromagnetic spectrum. (Redrawn from Government Accounting Office, *Research and Regulatory Efforts on Mobile Phone Health Issues,* GAO-01-545, General Accounting Office, Washington, DC, 2001.)

Exposure to electromagnetic radiation can result in health effects if the levels are high enough. Health effects are also determined by the frequency of the radiation. High frequency electromagnetic radiation (e.g., x-rays and gamma rays) have sufficient energy to cause ionization of atoms and molecules with which they interact. Exposure to ionizing radiation can cause serious biological damage, including cancer. RF radiations are at much lower frequencies and do not have sufficient energy to cause ionization (Figure 9.1). Nevertheless, RF radiation can produce biological effects by localized heating of tissues. Heating can kill cells and otherwise damage tissue. The microwave oven is an example of the useful application of heating by RF radiation in food preparation.

Mobile phones are designed to operate at power levels well below the threshold known to produce thermal effects. The question of health effects associated with mobile phone use has generally focused on the long-term or frequent use of low-power RF radiation that may produce nonthermal effects in tissues. However, there is little evidence that nonthermal related health effects occur. Most cell phones now on the market are limited to a maximum power level of 0.15 watts. This is less than the amount of power needed to light a flashlight bulb. By contrast, household microwave ovens use between 600 and 1100 watts of power.[2]

WILL CELL PHONES "FRY" YOUR BRAIN?

The major health effect of concern with cell phones is brain cancer. Currently there is no significant indication that chronic exposure to cell phone radiation has any detrimental effect on human health. Health effects from cell phone use are difficult to measure. The American Cancer Society estimates 18,500 new cases of brain (and other nervous system) cancer in 2005; of these, 12,800, or 70%, will be expected to die of their disease.[3] Although no biological, chemical, or physical agents have been clearly established as risk factors for brain cancer, several agents, including certain pesticides, herbicides and fertilizers, precious metals, and organic solvents, have been identified as possible carcinogens.

Studies of radiofrequency radiation as a cause of cancer are extensive. In the last five years, a number of comprehensive reviews of the scientific literature have been published. Studies have included exposures to a broad range of radiofrequency sources and include military personnel, electronic workers, medical workers, and industrial groups with high exposures. These reviews have found no evidence of a risk of cancer from exposure to radiofrequency radiation.[4]

Several epidemiological studies have focused specifically on the health effects of RF radiation from cell phones. The validity of conclusions in support of health effects in some large studies has been called into question because of serious study design weaknesses, including uncertainties in exposure assessment and the influence of reporting and other sources of bias. Collectively cell phone epidemiological studies have found no link between RF radiation from cell phones and brain cancer. This general conclusion is supported by results from laboratory animal studies.[5] Particular concerns have been raised regarding use of cell phones by children. Risks may be higher in children because the central nervous system is still developing in young children. Children are likely to live longer so as to express a health risk should such risks exist. A review of all of the available epidemiologic literature suggests that cell phone risks, if they exist, are highly uncertain and so small that they cannot be measured reliably.

Epidemiological studies are made difficult by the fact that cell phones are now used by about two-thirds of the population and brain and other central nervous system (CNS) cancers are rare forms of cancer.[6] If disease is distributed uniformly in the population, then about two-thirds of individuals with brain cancer (about 12,000 persons) would be expected to have a history of cell phone use. The high frequency of cell phone use among brain cancer patients makes it difficult to establish a causal relationship. However, causality can be excluded when the diagnosis of brain cancer is known to have preceded cell phone use.

Cell phone use rose dramatically in the 1990s (Figure 9.2) and early epidemiological studies may not have allowed for sufficient time to express cancer risk. However, studies now being conducted (in 2005) still do not show significant cancer risks. Sufficient time has now passed such that an effect would have been observed if there was one. Based on studies involving population exposure to ionizing radiation, the period for expression of health risk is probably a decade or more. Most studies have focused on short-term exposure of the entire body rather than on long-term exposure of the head that is characteristic of cell phone use. Furthermore, much of the research to date has investigated health effects of RF radiation at frequencies different from those used by cell phones. In addition most studies have focused on analog phone equipment. Now digital phones have become the standard technology. Digital phones transmit messages using discontinuous pulses, while analog equipment uses a continuous transmission mode. The available scientific literature does not demonstrate convincingly that the biological effects of RF exposure differ based on frequency or whether the signal is analog or digital.[7]

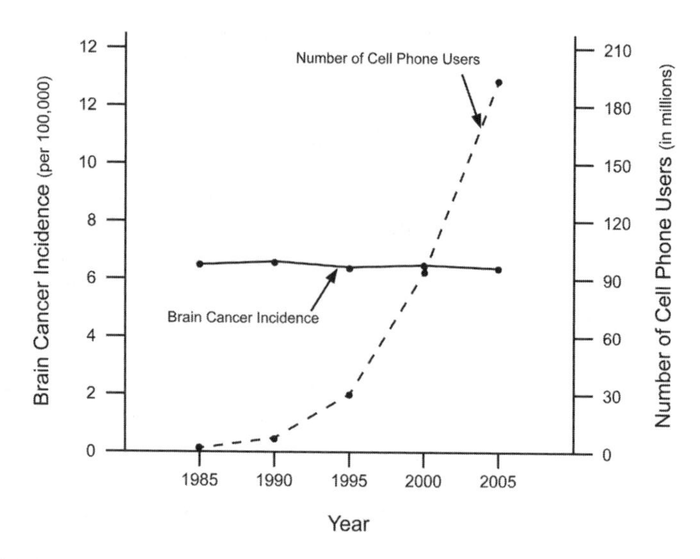

FIGURE 9.2 Cell phone use and brain cancer. Secular trend analysis suggests no relation between cell phone use and brain cancer incidence in the U.S. from 1985 to 2005. (Data for cell phone use is from the Cellular Telecommunications and Internet Association; brain cancer data are age-adjusted incidence rates from the Surveillance, Epidemiology and End Results (SEER) cancer database, U.S. National Cancer Institute.)

Secular trend analysis is a useful tool to determine whether there is any correlation between the incidence of brain cancer and cell phone use. Figure 9.2 compares incidence of brain cancer with growth in cell phone use from 1985 to 2005 and suggests that cell phone use is unrelated to brain cancer incidence. The secular trend analysis does not address the question of risk. The data simply suggest that either brain cancer is not correlated with cell phone use or if there is a correlation, it must be so small that it cannot be reliably measured due to the presence of other correlates that overwhelm the contribution of RF radiation. If cell phone use is a brain cancer risk factor, one might expect an increase in cancer incidence after the year 2000 as a consequence of the dramatic increase in cell phone use in the 1990s. No such increase in brain cancer has been observed.

Although devices such as cellular phones represent new technological developments, RF radiation in the form of radio and television equipment has been around for decades. No adverse health effects have been documented with the public uses of radio and television.

Epidemiological studies provide the most direct evidence of risk but alone are insufficient to fully establish causation. Several important statistical criteria must be met to establish causality, including strength of statistical association, the magnitude of the biological effect, and evidence of a dose-response relationship. There are also important biomedical criteria, including temporal relevance (exposure must precede appearance of the disease by an appropriate time frame consistent with biological understanding of the disease), evidence of a biological mechanism, and consistency of results across studies.[8] The statistical criteria may be evaluated in the context of epidemiological studies. The available evidence suggests that if a health risk exists, it is small and has not been observed consistently. No study to date has demonstrated a dose response relationship, i.e., that cancer risk increases with increased cell phone use. There is insufficient health risk data to draw conclusions about risks per unit time on the cell phone (i.e., does doubling time on the cell phone double the risk of cancer?). We do not know what the shape of the dose response curve is — if there is a meaningful dose-response relationship at all. Laboratory studies using cell culture or animal models can be used to evaluate biological mechanisms. A satisfactory mechanistic explanation of cell phone-induced cancer risk has yet to be proffered. One might expect that if RF radiation from cell phones is carcinogenic, the mechanism would involve damage to cellular DNA. There is little evidence from available cell and animal studies of genotoxic activity resulting from RF exposure.[9]

The recent debate on acoustic neuroma and cell phones illustrates how difficult it is to establish proof of risk. A small Swedish study reports increased risks of acoustic neuroma associated with cell phone use of at least 10 years duration.[10] The study reports that acoustic neuroma (a rare benign, noncancerous growth that arises from the vestibulocochlear nerve in the brain) appears to occur predominantly on the side of the head that is more frequently associated with cell phone use.[11] Health effects would be expected to predominate on the ipsilateral side of the head because radiation exposures are higher there. Further, risks were observed only for cell phone use in excess of 10 years. Individuals who used cell phones for less than 10 years did not show elevated risk. In contrast a Danish study found no correlation between pattern of cell phone use and frequency of acoustic neuroma. The results of this prospective population-based nationwide study that included a large number of long-term cell phone users suggest no radiogenic risk.[12] Currently there are few studies large enough to allow for a definitive conclusion about a long-term risk of acoustic neuroma. To confirm causality, further research is needed to elucidate mechanisms of radiogenic acoustic neuroma and to confirm the Swedish finding that brain lesions occur on the same side of the head as cell phone use. Whatever the risk, many people would consider the benefits of cell phones to be too important to abandon them. If the risk of acoustic neuroma turns out to be important, technological modifications in cell phones could be made to reduce exposures and thus reduce risks.

A recent court decision addresses the quality of the evidence in support of a causal link between cell phone use and brain cancer. In *Newman v. Motorola, Inc.* plaintiff Christopher Newman claimed that his use of a wireless hand-held telephone manufactured by Motorola, Inc. caused his brain cancer. After a substantial period of discovery and the designation of experts, both plaintiff and defendants (Motorola Inc. et al.) filed motions to exclude the others' proffered expert testimony. Judge Catherine Blake of the U.S. District Court in Baltimore, MD, rejected evidence offered by plaintiff's lawyers that alleged health risks from wireless telephones. On dismissing the case, Judge Blake ruled that plaintiff's witnesses presented "no sufficiently reliable and relevant scientific evidence" to support the claim of a link between wireless phone use and cancer. Motorola, Inc. and other defendants had argued that the scientific opinions presented by plaintiff's experts lacked scientific acceptance and failed to meet the legal test for expert testimony in such cases established by the U.S. Supreme Court. In agreeing, Judge Blake issued a significant decision that may have far reaching consequences in the evaluation of risk evidence.[13]

MANAGING PHANTOM RISKS

In the late 1990s, the U.K. Minister of Public Health convened an expert panel (the Stewart Group) to explore whether cell phones were a health hazard and, if so, what to do about it.[14] The Stewart Group was in response to a growing public, media, and government interest in the cell phone health issue. The Stewart inquiry was not established in response to any specific organized public concern, but clearly the government's mishandling of mad cow disease had sensitized the public to possible health risks. The cell phone controversy initially focused on placement of unattractive phone towers, but that may have been a smoke screen for the real concerns about possible health effects particularly in children. Media stories heightened public awareness and fueled the debate and public anxiety.

Despite scare stories in the press, people continued to use cell phones, and sales did not slack off, nor was there any scientific evidence of actual or potential harm. The government acted because it wanted to show it was being ultra-cautious and in control of a potential public health threat. This followed the government's inadequate response to mad cow disease and the safety of beef from bovine spongiform encephalopathy (BSE) infection.

The Stewart inquiry confirmed what was already known from numerous other studies — cell phones do not increase the risk of cancer. Nevertheless, it advocated caution and suggested that children should use cell phones as little as possible. The committee also recommended further research. The report alarmed the public and sent confusing messages. If there was no evidence of health risks, why was it necessary to recommend limited use by children? Were cell phones a risk or not? The Stewart report drew conclusions that went beyond the consensus view of other expert groups. Two issues in the Stewart report were particularly troublesome: first, the report stated that children are more susceptible to the effects of RF radiation because the CNS is still in development. Second, children absorb more energy from external electromagnetic fields than adults do. The Stewart report concluded that it is not possible at present to say that exposure to RF radiation, even at levels below national guidelines, is totally without adverse health effects. The level of uncertainty was considered sufficient to justify a precautionary approach. The Stewart report encouraged adoption of a precautionary approach until more detailed and scientific robust information on any health effects becomes available.[15]

The Stewart panel may have misunderstood key scientific knowledge or misinterpreted the available scientific data. Conclusions were made in the absence of evidence showing adverse effects of exposure on the CNS of children or adults, and without the identification of detrimental effects to which children may be more susceptible. The Health Council of the Netherlands concluded that the position of the Stewart group, which is based on the assumptions that developmental activity in the CNS and higher energy absorption characteristics in children's brain tissue enhance risk, is untenable. The Netherlands report concluded that there is no basis to recommend restricted use of cell phones by children.[16] The Stewart group's treatment of the physical aspects of RF technology was problematic regarding analysis of localized electromagnetic field intensities and densitometry to estimate exposure. The Stewart analysis may have resulted in an overestimation of energy absorption rates in children.[17]

The recommendation of the U.K. government to limit cell phone use by children raises important questions about how the precautionary principle is being applied as a risk-management tool. Should children be discriminated against because they might be at higher risk for health effects although evidence of higher risk is lacking? The U.K. government recognized that limited scientific evidence exists in support of health effects. Is the level of knowledge and understanding about cell phone risks sufficiently poor to warrant a precautionary approach? Recommendations of the Stewart report may be a reaction to the idea that public health decisions regarding technological risks cannot be solved by simply taking a probabilistic and rational approach. Nonquantifiable factors, including how risks and benefits are socially distributed (i.e., distributive justice), are also important considerations. Clearly, in this case, a precautionary approach is not justified. Radiological risks of cell phone use in children and adults have not been clearly established, and restricting cell phone use by children serves no useful purpose.

In 2004, the U.K.'s National Radiological Protection Board (NRPB) revisited the Stewart Group findings in response to a request from the government after the publication in May 2000 of the Stewart report. The NRPB found "no hard evidence" of any risk from the use of cell phones.[18] Yet the board supported the Stewart recommendation that children avoid using cell phones based on application of a "precautionary approach." The NRPB report says that the limitations of the research to date, and the fact that cell phones have not been in widespread use for a long time, mean that adverse health effects cannot be completely excluded. Again, this recommendation was not based on a finding of any actual risk to children, but rather was premised on the "risk of a risk." Banning cell phone use among children eliminates an important benefit of the technology. For example, cell phones afford parents peace of mind by staying in contact with their children.

The Stewart report and the subsequent NRPB report reflect the new politics of precaution that is now playing an ever-increasing and prominent role in European society. A precautionary approach to cell phone use might have been understandable if there was significant public concern or if there was real evidence of health risk, but neither was the case. The U.K. government directive to control cell phone use is an example of an arbitrary and capricious application of the precautionary principle.

IMPRUDENT PRECAUTION

The NRPB's decision to support the recommendations of the Stewart Group is indeed unfortunate. How much more evidence is needed about cell phone risks? Interestingly, by using precautionary measures, public decision makers may actually trigger concerns, distort and amplify cell phone risk perceptions, and lower trust in public health protection — consequences that are exactly opposite of what is intended.[19] The precautionary approach served to confirm the public sense that there are risks even though there is no evidence. The whole issue was badly managed by the U.K. government.

The precautionary principle is a risk-management strategy that states that when an activity or technology may harm human health or the environment, precautionary measures should be taken even if some cause-and-effect relationships are not fully established scientifically. While there have been some attempts to better define and operationalize the precautionary principle, most notably by the Commission of the European Communities (CEC), substantial ambiguity remains about the applicability and requirements of the precautionary principle.[20]

Because the precautionary principle does not prescribe courses of action regulators, industry and the public are left with little guidance to manage risk. Lack of guidance can lead to inappropriate applications of the precautionary principle.[21] The U.K. Department of Health recommendation to limit cell phone use by children is an egregious example.[22]

As discussed in Chapter 5, the CEC communication provides some guidance on when recourse to the precautionary principle is triggered. The Communication defines the precautionary principle as a risk-management tool, to be applied only after a scientific evaluation of the available risk data (i.e., risk assessment). The Communication describes two outputs from this risk assessment that are necessary to justify recourse to the precautionary principle. The risk assessment must identify potentially negative effects resulting from the product or activity, and the available scientific data must be so insufficient, inconclusive, or imprecise to make it impossible to "determine with sufficient certainty the risk in question." A political decision is then required to determine whether any precautionary action is appropriate, which is largely a function "of the risk level that is 'acceptable' to the society in which the risk is imposed."[23]

Clearly current understanding of radiological risks associated with cell phone use does not justify recourse to the precautionary principle. There is already an impressive body of epidemiologic and scientific data that supports the view that health risks are either zero or too small to be measured with confidence.[24] Furthermore, there is little evidence to support the view that children are at higher risk. Although there will always be scientific uncertainty associated with cell phone risk assessment, the epidemiologic and scientific data to date are of sufficient quality and quantity to support the view that radiological risks from cell phone use are either at or close to zero.

If the precautionary principle does apply to a risk, the CEC Communication describes a series of general principles that should govern application of the principle. First, the principle of *proportionality* should apply, which requires the chosen risk reduction measure to be proportional to the seriousness of the potential risk. It should include less restrictive alternatives that achieve the desired level of protection. Second, the principle of *nondiscrimination* requires that risk-reduction measures not be discriminatory in their application, and thus comparable situations should not be treated differently. Third, the principle of *consistency* requires risk reduction measures to be consistent with the measures already adopted in similar circumstances. Fourth, the *costs and benefits* of action and lack of action should be examined, including an economic cost/benefit analysis when this is appropriate and feasible. Fifth and finally, precautionary measures should be *provisional*, and should be reevaluated and if necessary modified with the development of scientific knowledge.[25]

Even if health concerns regarding cell phone use by children meet the triggers for the precautionary principle and recourse to a precautionary ban is justified, the U.K. directive fails to conform to at least three of the general principles of application of the precautionary principle as outlined by the CEC Communication. A limitation on cell phone use does not meet the *proportionality* test. There is no evidence that reducing cell phone usage results in a concomitant reduction in health risk. Without evidence of a dose-response relation, there can be no basis for support of a reduction in cell phone use to achieve a given level of radiological risk reduction. If the dose response were linear, a 5% reduction in cell phone use time would translate into a 5% reduction in health risk. However, there is little scientific evidence to support the view that health risk is related to time on a cell phone.

A recent Harvard study suggests that nonradiological risks may be far more significant than radiological ones. Banning cell phones while driving might reduce crash events by about 6% in the U.S. The percentage reduction in accidents is not known with certainty and may be as low as 2% or as high as 21%. Assuming a 6% diminution, risk reduction is substantial and translates into an annual prevention of 330,000 total traffic injuries, 12,000 serious injuries, and 2,600 traffic-related deaths.[26] Limiting cell phone use while driving is a more prudent risk reduction policy than banning cell phone use by children based on an unproven radiological risk.

Limiting cell phone use is also *discriminatory* and *inconsistent*. If a limitation on cell phone use is deemed appropriate, why is there not a limitation on computer use or loud music? These activities may also be associated with chronic health risks such as hearing loss. The health risk associated with cell phone use is no more certain, and is probably less certain, than these other risks.

The British cell phone debacle is only one of a number of health issues that has fallen prey to the precautionary principle. The precautionary principle is creating a wave of absurd and arbitrary risk decisions across Europe since the European Union adopted the principle into its laws in 1992.

The precautionary principle is triggered by the absence of scientific certainty in favor of safety. This is troubling because science can never be fully certain of risks or prove safety. The precautionary principle provides no limits on the application of precaution, which could be applied to any product or activity, thus leaving the principle prone to arbitrary application based on protectionist or other improper motives.

A recent analysis of more than 60 judicial decisions in the European Union (EU) court that cite the precautionary principle illustrates how arbitrary and capricious application of the precautionary principle can lead to imprudent risk-management decisions.[27] Despite relying on the precautionary principle to decide important cases, the EU courts failed to define or articulate the specific requirements or meaning of the precautionary principle. They simply invoked the precautionary principle as a wild card that justified whatever decision they reached.

Not surprisingly given its inherent ambiguity and lack of definition, the precautionary principle was applied by the EU courts in an arbitrary and capricious manner. In some cases, the courts acted quite sensibly to overturn regulatory decisions that lacked any scientific justification based on the precautionary principle. For example, the EU courts overturned decisions by the Netherlands to ban corn flakes because the added vitamins and nutrients could potentially harm an unusually susceptible person. They further overturned a decision by France to ban caffeinated energy drinks because the caffeine could potentially harm pregnant women. Likewise they overturned a decision by Denmark to ban cranberry juice drinks because the added vitamin C could potentially be harmful to some individuals who consume too much of the vitamin. All of these regulatory decisions were based on the precautionary principle, and all were rightly rejected by the EU courts as an unjustified departure from reasoned decision making.

In other cases, however, the courts applied the precautionary principle in an extreme and unreasonable manner. For example, in one case the EU court went on at great lengths about the importance of basing regulatory decisions on the most complete risk assessment possible. However, they then relied on the precautionary principle to uphold an EU decision to ban a certain antibiotic in animal food even though the pertinent risk assessment concluded that the antibiotic in question presented no foreseeable risks. The court said that as long as the EU relied on a risk assessment, it was free to come to its own differing conclusions. In other words, under the precautionary principle, the EU must base product bans on a risk assessment, but it makes no difference what that risk assessment actually says or concludes.

In still other cases, the EU courts dismissed the precautionary principle with little explanation as an insignificant and irrelevant provision that added nothing to the preexisting statutory requirements. In short, depending on the inclinations of the individual judges, and facts and interests involved, the precautionary principle can vary from absolutist, extreme measures mandating zero risk to a meaningless concept with no substantive effect on decisions.

Perhaps the most insightful statement in this growing body of EU case law was a warning by one jurist that the "precautionary principle has a future only to the extent that, far from opening the door wide to irrationality, it establishes itself as an aspect of the rational management of risks, designed not to achieve zero risk, which everything suggests does not exist." Based on the track record in Europe to date, including the most recent warning by the British advisory committee on cell phones, the precautionary principle has failed to establish itself as an aspect of the rational management of risks, and indeed has no future.

Regulations, advisories, or recommendations to reduce health risks related to cell phone use (or any health risk from a product or activity) should be established on sound scientific evidence and be based on appropriate socioeconomic and cost-effectiveness considerations. Incontrovertible proof of health risk or quantitative risk estimation with a high level of statistical certainty is not necessary; what is necessary is sufficient evidence to support the chosen risk reduction strategy. In the case of cell phone use by children, the available scientific evidence fails to support the view that reduced cell phone use will result in risk reduction. By focusing on radiological health risks, it appears that the U.K. government recommendation is misdirected. Nonradiological health risks are more important as suggested by the risk study of cell phone use while driving.[28] Children and the public in general would be better served by recommendations that focus on common sense strategies to reduce accidents associated with cell phone use. Prudent and reasonable uses of cell phones should be advocated.

They should include recommendations to limit cell phone activity while driving, riding a bicycle, or using equipment that requires the full attention of the operator. As many as 40 countries may restrict or prohibit cell phone use while driving. Since 2000, legislatures in every state in the U.S. have considered legislation related to cell phones and driving or driver distraction. Twenty-four states and the District of Columbia currently have laws concerning wireless phone use in the car.[29]

Although cell phone health risks appear to be a nonissue, additional research is warranted to understand more fully the nature of possible radiological risks. There is no need to invoke a precautionary position while pursuing additional information to reduce risk uncertainties. Research should include both epidemiological and laboratory-based studies. Although there is little evidence to suggest radiological risks will appear in the future, epidemiological studies emphasizing digital phone use in everyday practice should be encouraged for at least the next 10 to 20 years. Over this interval, radiological risks can be estimated with reduced statistical uncertainty. Laboratory studies to elucidate possible biological mechanisms should also be encouraged. Cancer is the primary radiological health effect of concern, and there is little biological understanding of carcinogenesis as a nonthermal process following RF exposure.

INTERNATIONAL CALLS

The U.K.'s answer to cell phone risks can only be described as extreme. In other countries like the U.S., France, and Germany, the problem is either below the public's radar screen or policy makers have dealt with it in a more moderate way.[30] Social perceptions of cell phone risks appear to be forged by specific national priorities and agendas. Sociopolitical ideologies and structures, environmentalism, and risk-management ideologies such as the precautionary principle are important determinants of national response to technological risks. Risk perceptions in the U.K. are no different than they are in other countries. But media and political concerns appear to be much higher.

The reaction of government authorities determines the extent to which the cell phone health threat becomes a national issue. In the U.K. the Stewart report was government-initiated and had nothing to do with any organized public concern about health risks of cell phones. But the report raised public concerns and anxiety because of confusing messages about whether risks are real and made ill-advised recommendations to limit phone use by children. In the U.S. the cell phone debate is at best a local issue and focuses on placement of cell phone towers, particularly in school districts and related health questions from school boards. The issue never became nationalized because of the lack of convincing scientific evidence of harm. Compared with many European countries, local coverage of cell phone health risks is relatively subdued in the U.S. where wireless operators have worked closely with local authorities and citizen's groups to establish uniform standards over mast sitings. They have built greater trust and cooperation in an effort to reduce conflict. By contrast, in Italy, Switzerland, and Australia sustained concerns about health effects of electromagnetic fields at the local level have led to the high-profile emergence of radiation risks as a national issue. The success of the localized efforts in Italy can be traced to an Italian political framework characterized by strong regional authority.

Following the release of the Stewart report, several countries including the Netherlands, France, and Canada, commissioned their own reviews. All came essentially to the same conclusion of no harm. Nevertheless concerns have not gone away. Cell phone hazards have been linked to possible health effects from other sources of electromagnetic fields, including power lines. In the U.S. concerns over possible health effects from cell phones derived from reports of childhood leukemia following exposure to electromagnetic fields near power lines in the late 1970s. The possibility that low-frequency electromagnetic radiation could cause cancer never entirely disappeared, even though the initial findings of harm in the 1970s studies could not be confirmed. Although Italians have enthusiastically embraced cell phone technology, underlying concerns about risks persist because of early reactions to microwave facilities such as radar installations, power lines, and TV transmitters. General concerns about hazards from electromagnetic fields in Australia have also fueled questions about cell phone risks. Neither cell phones nor towers have been the subject of significant concern in France. But the French government conducted a careful review of risks to publicly demonstrate its commitment to consumer safety even though no specific concerns about cell phone risks had been raised.

Environmentalism is a major driver of public concern and national response to cell phone risks. Environmentalism is much stronger in Europe than it is in the U.S. In Europe cell towers or masts appear to be the focus of attention, not cell phones themselves. Towers impact the environment and when sited near schools raise the possibility of risks for children. Impact of cell towers, particularly from an aesthetic perspective, became important in the 1990s. Local antiradiation efforts in the U.S. have generally been unsuccessful in drawing support from environmental organizations as has been done in Europe. Environmental groups like the Sierra Club have largely stayed away from the issue. In Italy electromagnetic fields are now considered a source of environmental pollution. "Electrosmog" is a routine expression in Italian society and refers to the sea of electromagnetic radiation from broadcast and mobile telephone transmitters that everyone is exposed to. The "electrosmog" issue has enjoyed a high media profile and even includes a dispute between the Italian government and the Vatican over the intensity of the signal from Vatican radio.

The strong sense of environmentalism in Europe has led to precautionary approaches to risk management. Although cell phone risks are difficult to measure, several European nations have taken a precautionary approach to the management of cell phone technology, including design, placement, and operations of towers and other telecommunications equipment. Italy and Switzerland have pursued similar policies of extreme caution with regard to electromagnetic field emissions. Regulations are highly restrictive and at odds with international recommendations. The clear precautionary posture in the U.K. directive to limit cell phone use by children may have been a knee-jerk reaction to the government's handling of mad cow disease. The questionable government practices in managing mad cow disease may have led to the government going overboard in providing protection from cell phones.

In spite of some government policies advocating extreme caution, the public does not consider cell phones to be hazardous. Cell phone purchases continue to rise. In the U.S. there was never a grassroots effort to change the landscape of the cell phone industry. Isolated school boards and other groups have argued about placement of cell phone towers particularly in areas where children congregated. However, there has never been a national effort to curb the technology. The tragic shootings at Columbine High School near Denver, CO, in 1999 and the terrorist acts that took thousands of lives in New York in September 2001 illustrate that cell phones serve a useful purpose and are a valuable commodity with redeeming social worth outweighing any possible health risk. The size and complexity of American society works against any grassroots nationalized effort to moderate cell phone. There could not have been the sort of newspaper campaign against cell phones seen in the U.K., for example, because the U.S. lacks a widely read national press.

The cell phone scare in the U.K. is a good lesson on how not to react to concerns about emerging technologies. And there is plenty of blame to go around. The print and broadcast media in the U.K. broadcast sensational stories that served to heighten public concern over essentially phantom risks. The government overreacted in part as a response to its mismanagement of mad cow disease. Scientists were unclear about what the health risk message should be. Scientists were right to indicate that knowledge of health risks is uncertain but the emphasis on what we do not know as compared to what we do know led to confusing public messages.

The cell phone controversy in the U.K. is an example of how risk assessment and risk management can be mishandled in the public arena. There is a need for clear communication. Conveying uncertainties is an important component in message crafting but should not be used to avoid taking a firm, responsible position on a controversial issue. When it is clear that health risks are too small to measure, reliably calling for more research may reduce risk uncertainties but it does not change the essence of the message. There is no value in saying that we don't really know the risks and that our information is highly uncertain when the science doesn't really say that.

NOTES AND REFERENCES

1. Government Accounting Office, *Research and Regulatory Efforts on Mobile Phone Health Issues,* GAO-01-545, General Accounting Office, Washington, DC, 2001.
2. Ibid.
3. American Cancer Society, *Cancer Facts and Figures 2005,* American Cancer Society, Inc., Atlanta, GA, 2005.
4. Elwood, J.M., A critical review of epidemiologic studies of radiofrequency exposure and human cancer, *Environmental Health Perspectives,* 107 (Suppl. 1), 155, 1999; Moulder, J.E. et al., Cell phones and cancer: what is the evidence for a connection?, *Radiation Research,* 151, 513, 1999; Independent Expert Group on Mobile Phones (IEGMP), *Mobile Phones and Health,* Independent Expert Group on Mobile Phones, c/o National Radiological Protection Board, Chilton, (Stewart Report), 1999; Health Council of the Netherlands, *Mobile Telephones: An Evaluation of Health Effects,* Publication no. 2002/01E, Health Council of the Netherlands, The Hague, 2002; Boice, J.D. and McLaughlin, J.K., Epidemiological studies of cellular telephones and cancer risk — a review, Report number 2002: 16, Swedish Radiation Protection Authority, Stockholm, 2002; Royal Society of Canada, *A Review of the Potential Health Risks of Radiofrequency Fields from Wireless Telecommunication Devices,* RSC.EPR 99-1, The Royal Society of Canada, Ottawa, 1999.

5. Hardell, L. et al., Use of cellular telephones and the risk for brain tumors: a case-control study, *International Journal of Oncology,* 15, 113, 1999; Hardell, L. et al., Cellular and cordless telephones and the risk for brain tumors, *European Journal of Cancer Prevention,* 11, 377, 2002; Inskip, P.D. et al., Cellular-telephone use and brain tumors, *New England Journal of Medicine,* 344, 79, 2001; Johansson, C. et al., Cellular telephones and cancer — a nationwide cohort study in Denmark, *Journal of the National Cancer Institute,* 93: 203, 2001; Health Council of the Netherlands, *Mobile Telephones: An Evaluation of Health Effects*, Publication no. 2002/01E, Health Council of the Netherlands, The Hague, 2002; Boice, J.D. and McLaughlin, J.K., Epidemiological studies of cellular telephones and cancer risk — a review, Report number 2002: 16, Swedish Radiation Protection Authority, Stockholm, 2002.
6. Brain and other central nervous system cancers are relatively rare. According to the American Cancer Society, 173,000 lung cancers, 213,000 breast cancers, 232,000 prostate cancer and 145,000 colon and rectum cancers are estimated for 2005 compared to 18,500 brain and other central nervous system cancers. See American Cancer Society, *Cancer Facts and Figures 2005,* American Cancer Society, Inc., Atlanta, GA, 2005.
7. *Supra* note 1.
8. Hill, A.B., The environment and disease: association or causation?, *Proceedings of the Royal Society of Medicine,* 58, 295, 1965.
9. Moulder, J.E., Cell phones and cancer: what is the evidence for a connection?, *Radiation Research,* 151, 513, 1999.
10. Lonn, S. et al., Mobile phone use and the risk of acoustic neuroma, *Epidemiology,* 15, 653, 2004; A commentary by Savitz emphasizes the importance of the Lonn study and the need to interpret the risk with caution. See Savitz, D.A., Mixed signals on cell phones and cancer, *Epidemiology,* 15, 651, 2004.
11. Information on patterns of cell phone use was based on subject recall in the Lonn study (*supra* note 11). Recall bias can be a serious problem because subjects recall information from memory. Subjects with acoustic neuroma may be more likely to associate cell phone use with the side of the head with disease.
12. Christensen, H.C. et al., Cellular telephone use and risk of acoustic neuroma, *American Journal of Epidemiology,* 159(3), 277, 2004.
13. *Newman v. Motorola, Inc.*, 218 F. Supp. 2d 783 (D. Md. 2002).
14. Independent Expert Group on Mobile Phones (IEGMP), *Mobile Phones and Health,* Chilton, U.K., Independent Expert Group on Mobile Phones, c/o National Radiological Protection Board, Chilton, (Stewart Report), 1999; U.K. Department of Health, Government Response to the Report from the Independent Expert Group on Mobile Phones (Stewart Group), Department of Health, London, 2002.
15. Balzano, Q. and Sheppard, A.R., The influence of the precautionary principle on science-based decision-making: questionable applications to risks of radiofrequency fields, *Journal of Risk Research,* 5, 351, 2002.
16. Health Council of the Netherlands, *Mobile Telephones: An Evaluation of Health Effects,* Publication no. 2002/01E, Health Council of the Netherlands, The Hague, 2002.
17. *Supra* note 15.
18. National Radiological Protection Board (NRPB), *Mobile Phones and Health 2004.* Documents of the NRPB, Volume 15, Number 5, NRPB, Chilton, 2004.
19. Wiedemann, P.M. and Schutz, H., The precautionary principle and risk perception: Experimental studies in the EMF area, *Environmental Health Perspectives,* 113, 402, April 2005.

20. Commission of the European Communities (CEC), *Communication from the Commission on the Precautionary Principle*, CEC, Brussels, 2000.
21. Mossman, K.L. and Marchant, G.E., The precautionary principle and radiation protection, *Risk: Health, Safety & Environment,* 13, 137, 2002.
22. U.K. Department of Health. *Supra* note 14.
23. *Supra* note 20.
24. *Supra* note 4.
25. *Supra* note 20.
26. Cohen, J.T. and Graham, J.D., A revised economic analysis of restrictions on the use of cell phones while driving, *Risk Analysis,* 23(1), 5, 2003.
27. Marchant, G.E. and Mossman, K.L., *Arbitrary & Capricious: The Precautionary Principle in the European Union Courts,* AEI Press, Washington, DC, 2004. Professor Gary Marchant (Arizona State University College of Law) provided the brief legal descriptions of the EU court cases mentioned in this section of the chapter.
28. *Supra* note 26.
29. National Conference of State Legislatures, *Cell Phones and Highway Safety 2005 State Legislative Update,* National Conference of State Legislatures, Denver, CO, 2005.
30. Much of this section is derived from Burgess, A., *Cellular Phones, Public Fears, and a Culture of Precaution,* Cambridge University Press, Cambridge, 2004.

10 PR Campaign: Proportion, Prioritization, and Precaution

The impressive safety record of the nuclear industry (including but not limited to medical uses of radiation and nuclear power) is *prima facie* evidence that the current system of radiation protection works. Regulatory limits, particularly for the public, are set well below observable risk levels. Radiation doses encountered in occupational settings under normal operating conditions are also tiny fractions of levels known to produce acute injury. Risks of radiogenic cancer are so low that they are very difficult to measure, if they can be measured at all. Despite the stellar worker and public safety record, the radiation protection framework is complex and confusing, promotes public fear of low doses of radiation, and is increasingly expensive. A risk-based system of protection is difficult to defend because of large uncertainties in risk. Communicating risk information is challenging because the public has difficulty understanding small risks.

The current radiation protection framework needs restructuring to simplify protection and improve public communication. Central to reorganization is the switch from a risk-based to a dose-based system of protection. Dose proportion is an effective communication tool, and quantifying dose reduction is a meaningful metric for protection because dose can be measured directly. Further radiological protection must focus on radiological risks within the broader context of competing risks. Prioritization of risk is the key to efficient public health protection. Prioritization does not marginalize management of small risks but recognizes that the greatest gains in health protection are realized by managing large risks. The public continues to push towards a zero risk tolerance in which the balanced cost-benefit approach of the as low as reasonably achievable (ALARA) philosophy is replaced by a precautionary approach of better safe than sorry. The view is that risks should be avoided at all costs. Certainly some precaution is prudent and necessary in risk management given that scientific certainty is unattainable. But in radiation safety, strict precaution is not required. Sufficient information about health risks is known at radiation levels found in occupational and environmental settings.

Proportion, prioritization, and precaution is the blueprint for a new radiological protection system. This PR campaign leads to a more efficient system that will maintain the high level of protection currently enjoyed in the nuclear sector but is easier to communicate, less complex, and more cost effective.

PROPORTION

The "dose proportion" concept was introduced in Chapter 7 as an alternative to risk quantification at low dose. The problems associated with assessing very uncertain risk using predictive theories like LNT are avoided by using dose proportions. Quantification of small risks is a problem because the science underlying the risk-assessment process is uncertain. The public has great difficulty comprehending very small probabilities. Furthermore, the public mistakenly concludes that theory-derived risks are "real" when, in fact, they are nothing more than "speculates."

Dose is preferable to risk as an indicator of health detriment because it can be measured directly. Risk is inherently more uncertain than dose because risk is a dose derivative. However, even dose measurements must be interpreted cautiously because of new scientific understanding suggesting that very small doses may not relate to health detriment.

The dose proportion is the ratio of the measured or calculated dose to a reference dose. The dose proportion is dimensionless and thus independent of specific dose units. Dealing with proportions (rather than risk probabilities) is a more effective way of communication because the numbers have meaning, provided that the reference dose is clearly explained. The dose proportion should also include some expression of variability and uncertainty. Variability has to do with changes in the value due to natural tendencies or differences in the environment (such as geographical differences in natural background radiation levels); uncertainty has to do with the statistical or random errors in measurement.

Dose proportions for selected environmental, occupational, unplanned (accident), and medical exposure scenarios are listed in Tables 10.1 and 10.2.[1] Values can vary considerably. A dose proportion equivalent to unity means that the dose received is equal to one year's worth of natural background radiation (about 1 mSv per year

TABLE 10.1
Dose Proportions for Selected Environmental, Occupational, and Accident Exposures

Exposure	Dose Proportion
X-ray security screening	0.0003
Transcontinental Airline flight	0.1–0.5
Chernobyl accident	14
Hypothetical nuclear terrorism incident	3–30
Nuclear worker average annual dose	20
Annual exposure on international space station	170

Sources: Values in the table were calculated from dose estimates provided in: National Research Council, *Airline Passenger Security Screening: New Technologies and Implementation Issues,* National Academy Press, Washington, DC, 1996; Brenner, D.J. et al., Cancer risks attributable to low doses of ionizing radiation: Assessing what we really know, *Proceedings of the National Academy of Sciences,* 100, 13761, 2003; Cardis, E. et al., Risk of cancer after low doses of ionizing radiation: retrospective cohort study in 15 countries, *British Medical Journal,* 331, 77, 2005.

TABLE 10.2
Dose Proportions for Selected Medical Exposures

Exposure	Dose Proportion
Chest x-ray	0.2
Mammogram	3
Pediatric CT	25
Fluoroscopy (1 minute)	< 100
Cardiac catheterization	100
Radiation therapy for cancer	> 30,000

Sources: Values in the tables were calculated from dose estimates provided in: National Council on Radiation Protection and Measurements (NCRP), *Exposure of the U.S. Population from Diagnostic Medical Radiation*, NCRP Report No. 100, NCRP, Bethesda, MD, 1989; Brenner, D.J. et al., Cancer risks attributable to low doses of ionizing radiation: Assessing what we really know, *Proceedings of the National Academy of Sciences*, 100, 13761, 2003.

excluding contributions from radon gas). Dose proportions less than or more than unity represent submultiples or multiples of the reference natural background radiation level respectively. None of the dose proportions listed in Table 10.1 is associated with deterministic effects such as skin reddening or bone marrow depression. Certain high dose medical procedures such as radiotherapy for cancer produce acute effects (Table 10.2). Dose proportions in excess of about 500 would be needed to observe deterministic effects. Risks of cancer are increased for dose proportions greater than 100. Radiogenic cancer risks have been difficult to observe for dose proportions below 100, so any risk estimates must be viewed with considerable caution.

For nonmedical exposures, dose proportions range over 6 orders of magnitude (Table 10.1). Environmental exposures are routinely very small. Airline passengers must undergo security screening at airports, and the x-ray dose is a tiny fraction of natural background radiation levels. Passengers on a round trip airline flight from New York to London are exposed to a small additional dose of radiation (about 0.1 to 0.5 mSv) due to increased cosmic ray exposure at high altitude.

Accidental and occupational exposures by their nature are highly variable. The dose proportion for the Chernobyl accident reflects a calculated average whole body dose of 14 mSv over a 70-year period to 500,000 individuals in rural Ukraine in the vicinity of the accident site. Some Chernobyl victims received doses much higher or lower than the average depending on their location relative to the plant. The dose proportion for the hypothetical terrorist event (Table 10.1) is determined theoretically and is highly uncertain. It is based on computer generated dose distributions over a 20-block radius from a hypothetical nuclear terrorism incident involving dispersal of radioactive cesium from a radiological dispersal device or “dirty bomb” detonated in a large urban environment. Workers in the nuclear industries may be exposed to large doses unless proper controls are in place. In several studies of health effects in nuclear workers, average doses were estimated to be about 20 mSv. Exposures to workers on the international space station are very high because of intense cosmic radiation from outer space. On Earth the atmosphere filters out most of the cosmic rays.

Medical and dental x-rays are the largest anthropogenic source of radiation exposure and account for about 18% of the total annual radiation exposure to the U.S. population.[2] Diagnostic studies to obtain important medical imaging information usually involve small doses, but in some studies the dose can be quite high (Table 10.2). Cancer risks from diagnostic studies are too small to measure and are considered negligible in the context of expected medical benefits. A chest x-ray results in a skin entrance dose of about 0.2 mSv.[3] A screening mammogram results in a breast dose of about 3 mSv. These doses are very small compared to doses received by children during computerized tomography (CT) scans of the abdomen. Cardiac catheterization is an interventional procedure that can result in large doses to the patient in order to assess patency of coronary arteries. Unlike other applications of radiation in medicine, very large doses are needed in cancer therapy to kill cancer cells. Doses in radiation therapy and in complicated interventional cases are large enough to produce deterministic effects.

A scale of dose proportions can be readily calibrated by comparing values to specific benchmarks such as dose limits and doses known to produce stochastic and deterministic effects. Benchmarks do not assume any particular dose response but are used to put measured dose proportions in perspective. In the context of the dose proportions listed in Table 10.1, dose proportion representing the maximum allowable annual radiation exposure to workers is 50; a dose proportion of 100 represents the minimum dose associated with elevated cancer risk; the threshold for deterministic effects is represented by a dose proportion of 500.

There is little epidemiological evidence in support of statistically significant cancer risks for dose proportions up to 100. Environmental, occupational, and most diagnostic medical exposures are in this range. Risks are highly uncertain, and the shape of the dose-response function is unclear. Accordingly, dose proportions below 100 must be interpreted with caution.

Dose proportions are important in communicating health risks to the public. Risks are interpreted differently by experts and by the public. Scientists, engineers, and other experts prefer to use quantitative approaches. However, the public is generally ill prepared to deal with risk numbers, particularly if they are expressed as percentages or ratios that can be easily misinterpreted. Accordingly, quantification of small risks should be avoided. Qualitative expressions are more meaningful (e.g., "The risk is considered insignificant" or "The risk is so small that health impacts are considered unlikely"). Use of qualitative descriptions (e.g., high, moderate, low) or other semi-quantitative approaches can be effective communication tools. Use of qualitative expressions of health detriment for dose proportions up to 100 is preferred because it is a scientifically defensible approach. Because of the large uncertainties in small risks, quantification has little meaning and may reflect a degree of confidence not supportable by the data. The public is more likely to understand qualitative expressions of risk even though quantifying risk would appear to be the superior approach.

Dose proportions may be particularly useful in the informed consent process to ensure that risks of medical radiation exposure are stated accurately and in language understandable to the patient or research subject. Patients and research subjects should understand the risks before consenting to undergo medical diagnostic procedures or therapy. Dose proportions place radiation doses in perspective. For example, assume a patient undergoes a mammogram in which the midline breast dose is 3 mSv (equivalent to a dose proportion of 3 as shown in Table 10.1). The following statement may be included in the informed consent form to help the patient better understand the significance of the dose:[4]

> The amount of radiation received in this procedure is equivalent to the amount of naturally occurring background radiation that all individuals in the United States receive in a three-year period. The risk from an exposure of this magnitude is generally considered negligible.

In cases when radiation doses are high (e.g., doses greater than 200 mSv), quantitative expressions of risk are warranted as long as uncertainties in risk estimates are provided. Risks of radiogenic cancer have been observed directly for doses in excess of 200 mSv, and LNT theory provides reasonable and conservative estimates of risk. Qualitative and quantitative descriptions of risk carry different but equally valuable perspectives. Each approach has its own biases and limitations.

PRIORITIZATION

The primary goal of public health protection is the control of environmental factors known to cause disease. The major risk factors compromising U.S. public health are well known. Smoking, diet, and lack of exercise are now recognized as the major causes of death in the U.S. Cigarette smoking is responsible for 30% of cancer deaths and is also a major contributor to cardiovascular mortality. Obesity is increasing at an alarming rate, especially in developed countries, with major adverse consequences for human health. The U.S. Centers for Disease Control and Prevention (CDC) reports that from 1988–1994, 60% of adults in the U.S. were overweight or obese, a figure that rose to 65% in 1999 to 2000. Obesity is considered to be a contributing factor to chronic diseases such as heart disease, cancer, and diabetes. According to CDC, physical inactivity now ranks as the number three cause of death in the U.S. Lack of exercise plays a role in many chronic diseases that lead to early mortality. Control of these risk factors is challenging. Any serious risk-management initiative will depend on individual and community programs that focus on modifying personal behaviors and public education through government, industry, and commercial messages.

Numerous epidemiological studies on geographical and temporal variations in cancer incidence, as well as studies of migrant populations and their descendants that acquire the pattern of cancer risk of their new country, indicate that over 80% of cancer deaths in Western industrial countries can be attributed to factors such as tobacco, alcohol, diet, infections, and occupational exposures. Diet and tobacco together account for about two-thirds of cancer deaths. In a recent compilation of data, it was estimated that about 75% of all cancer deaths in smokers and 50% of all cancer deaths in nonsmokers in the U.S. could be avoided by elimination of these risk factors.[5]

"Cancer" represents over 100 different diseases. For some cancers environmental risk factors are well known (e.g., smoking and lung cancer); for other cancers environmental causation is poorly understood (e.g., prostate cancer). Known human carcinogens do not increase cancer risks uniformly. Whole-body exposure to ionizing radiation increases cancer risks in some tissues (e.g., thyroid gland) but has no effect on other tissues (e.g., prostate gland). Cigarette smoking clearly increases cancer of the lung and upper aero-digestive tract but is not known to affect certain other tissues such as the thyroid gland or brain.

Which cancer risk factors are important and which ones are insignificant? How are risk factors perceived, and what does perception have to do with risk prioritization? Quantitative measures of risk are not necessarily congruent with public perceptions. More than one-third of Americans believe Acquired Immune Deficiency Syndrome (AIDS) is the most urgent health problem facing the world today, ranking it second to cancer; yet concern about AIDS has been decreasing. Smoking is recognized as the single most important preventable risk factor. But tobacco-related cancers (e.g., lung cancer) do not receive the scientific support that other less lethal cancers do (breast, prostate). Research funding seems out of sync with mortality risks.[6]

Unfortunately, we fear the wrong things and spend money to protect ourselves from minor risks. We allocate limited resources to manage risks that pose little, if any, health concerns and ignore larger risks that have a significant health impact. The prioritization problem is illustrated in the U.S. Environmental Protection Agency's (EPA) recent tightening of the arsenic standard for drinking water. The EPA revised the current drinking water standard for arsenic from 50 parts per billion (ppb) to 10 ppb.[7] The agency claims that the more restrictive limit will provide additional health protection against cancer and other health problems, including cardiovascular disease and diabetes. To meet the new arsenic standard, the public is spending millions of dollars on new water mains and water treatment plants. Water rates across the country have increased significantly to meet the new EPA guidelines. But there is little evidence that reducing the standard from 50 ppb to 10 ppb has any direct benefit on the public health. In the meantime efforts to curb poor air quality in large urban environments remain seriously underfunded. Poor air quality has been clearly linked to ill health.

Small risks should not be ignored but must be placed in perspective. It makes little sense to allocate significant resources to management of risks that have little public health impact when larger, more significant risks remain poorly controlled. The key word is "significant." What may be a minor risk to some may be considered significant by others. Small individual risks should be managed if appropriate control technologies are available with due consideration for competing risks and cost-benefit analysis.

This book has promoted the idea that analyzing and discussing individual risks without regard to the presence of other risks is inappropriate. Often risks that may appear to be important when considered in isolation may become less significant when compared to other risks in the environmental or occupational setting. An individual risk cannot be evaluated in isolation but only in relation to, and with, other risks. To understand what risk (and dose) means requires that it be placed in appropriate context, including comparisons of dose and comparisons of health outcomes.

The fact that most radiological risks are small and cannot be reliably measured does not minimize their importance. We should manage such risks given the availability of resources. Major public health gains in the cancer wars are achieved by focusing on diet and smoking. Controlling radiological risks is a small battle but nevertheless worth fighting because we have the technological capabilities of doing so. If we can avoid exposure, we should do so. Individuals who might be exposed to radiation or other carcinogens should be protected. Although a utilitarian philosophy is the most cost-effective approach to public health protection, ideally no one should be left unprotected. Of course, resources are severely limited even in the wealthiest countries, and unfortunately, some people cannot be protected to the extent that we would like. Public officials and risk managers are left with the difficult task of deciding who can be adequately protected and who cannot.

PRECAUTION

The precautionary principle is an extremist approach to risk management that advocates zero risk tolerance. In radiological protection there is no justification for implementation of a precautionary risk-management approach.[8] As discussed in the case study on cell phones (Chapter 9), the precautionary principle has become the vehicle for introducing arbitrary public policy and regulations. Zero risk tolerance imposes unreasonable restrictions on technologies such that little or no progress is possible. Holding up technological progress because of concerns about proving safety beyond a reasonable doubt is a way of guaranteeing that advances will be significantly and unjustifiably hindered.

The precautionary principle is set within a consequential framework in which disproportionate weight is placed on probabilities, however small, of disastrous outcomes. Rather than a simple net utility rule (i.e., balancing risk against benefit), the precautionary principle proposes that even if there is a favorable weighing of costs against benefits the technology should nevertheless be rejected if even very small risks are large enough. The cell phone debate in the U.K. clearly illustrated this.

Recourse to the precautionary principle is used by decision makers to cope with public fears when technological risks are considered serious and cannot be excluded. The precautionary principle sounds relatively innocuous on its face, giving regulators broad authority to err on the side of safety by adopting temporary and proportionate measures to prevent any serious and irreversible harm to human health or the environment. In practice, however, the precautionary principle is creating a wave of absurd and arbitrary risk decisions across Europe since the European Union adopted the principle into its laws in 1992.[9]

Further, it is unclear whether implementation of the precautionary principle actually accomplishes what is intended. The objective of the principle is to reassure the public by taking extreme protective actions by removal of or severe limitation of the technology. However, such actions may have the opposite effect by alarming the public. Precautionary action may be interpreted as recognition by authorities that the technological risk is serious and the public needs proportional protection. Whether the public feels reassured or alarmed by precautionary decisions will depend on the initial level of public concern, the extent of public consultation, and the level of trust between the public and authorities.[10]

Precautionary approaches are reasonable for technologies in which activities or products are known to produce very serious risks and little benefit. The precautionary principle has an important role to play in cases such as global warming where potential consequences are severe and eliminating certain technologies would diminish the risk. In this case eliminating technologies may avoid serious catastrophe. But there is little justification for a precautionary approach in the management of most technological risks. The precautionary principle avoids the rational analysis of weighing costs against benefits in decision making and can lead to potentially serious risk trade-offs and unintended consequences that may be more serious than the risks the precautionary principle was intended to avoid.

The ban on dichloro-diphenyl-trichloroethane (DDT) in the 1960s is an example of how misapplication of the precautionary principle can lead to tragic consequences. DDT is a highly effective pesticide for the control of a variety of insects including mosquitoes and was a welcome substitute for toxic insecticides containing arsenic, mercury, or lead. DDT was introduced as a pesticide in the 1940s but was banned in the U.S. and other countries in the 1960s because of questions of safety. The DDT ban led to a dramatic increase in malaria because mosquito populations could no longer be effectively controlled. Experiences in Sri Lanka testify to DDT's effectiveness. In the late 1940s, prior to the introduction of DDT, about 3 million cases of malaria were reported in Sri Lanka. After introduction of DDT in 1963, the number of malaria cases plummeted to fewer than 100. Malaria incidence remained low until the DDT ban was introduced in the late 1960s. By 1968 1 million new cases of malaria were reported and in 1969 the number of malaria cases reached 2.5 million, roughly the same number as pre-DDT levels. The decision to ban DDT was based in part on the belief that DDT is a human carcinogen. However, the weight of scientific evidence suggested otherwise; by 1970 the overwhelming scientific evidence was that DDT is both safe and effective. DDT saved lives, but unsubstantiated concerns about safety led to an unwarranted ban resulting in death and disease in millions of people.[11]

On February 28, 2005, the French Parliament finalized amendments to its national Constitution to incorporate an Environmental Charter. Among other provisions, this charter mandates the application of the precautionary principle to all regulatory decisions in France. This action to enshrine the precautionary principle in its Constitution commits France to a path that will severely damage both the economy and public health of France and its trading partners.

In its current form, the precautionary principle is not sustainable in the long run. If the principle has a future it must be reformulated to establish itself as an aspect of the rational management of risks. It is not designed to achieve zero risk, which everyone agrees cannot be achieved. Based on the Europe experience, the precautionary principle has shown itself to be a reckless, unreasonable, and irresponsible experiment. The recent actions of the French Parliament ensure that the current version of the precautionary principle will inflict a lot more harm and mischief around the world before it eventually and inevitably collapses upon itself.[12]

The current system of radiological protection promotes the idea that any dose of radiation is potentially harmful. Coupled with the notion that even the tiniest amounts of radiation can be reliably detected, it is easy to understand how the public concludes that even the smallest measurable dose is dangerous. In the public's view the goal of any risk-management system should be to clean up every radioactive atom whether natural or anthropogenic. A system of protection that promotes these ideas does a disservice to the public. Although the current framework adequately protects workers and the public, it is overly complex and fails to distinguish real from theoretical risks.

The underlying theme in this PR campaign is simplicity. Simplicity focuses on identifying essential elements of a system of protection. Measurement and reduction in dose is the essence of a system of protection; risk cannot be measured at dose levels encountered in most environmental and occupational settings. Risk loses meaning when probabilities are very small and beyond everyday experiences; risk cannot be readily understood by the public. Simplicity focuses on similarities between comparable entities like anthropogenic and natural background radiation. The only property that distinguishes natural and anthropogenic radiation is its source. Ionizing radiations, regardless of its sources, interact with matter in the same way and produce the same spectrum of biological effects. It is this principle that makes dose comparison from natural and man-made exposures valuable as a way of putting dose into perspective.

NOTES AND REFERENCES

1. Dose proportions were calculated using annual natural background radiation (excluding contributions from radon gas) as the reference. Values shown will increase or decrease depending on use of other reference doses (see, for example, dose proportions for residential radon in Chapter 8).
2. National Research Council, *Health Risks from Exposure to Low Levels of Ionizing Radiation,* BEIR VII Report, National Academies Press, Washington, DC, 2005.
3. Medical doses are usually expressed as absorbed doses using units of gray (Gy) or mGy (1 Gy = 1000 mGy). For x-rays and gamma rays 1 Sv = 1 Gy.
4. For this example natural background radiation levels are assumed to be 1 mSv per year excluding contributions from radon. See Mossman, K.L., Radiation protection of human subjects in research protocols, in Mossman, K.L. and Mills, W.A. (Eds.), *The Biological Basis of Radiation Protection Practice,* Williams & Wilkins, Baltimore, MD, 1992, 255.
5. Luch, A., Nature and nurture — Lessons from chemical carcinogenesis, *Nature Reviews Cancer,* 5, 113, 2005.
6. Jaffe, H., Whatever happened to the U.S. AIDS epidemic?, *Science*, 305, 1243, August 27, 2004; Dennis, P.A., Disparities in cancer funding, *Science,* 305, 1401, September 3, 2004.
7. The Safe Drinking Water Act requires the EPA to revise the existing 50 parts per billion (ppb) standard for arsenic in drinking water. On January 22, 2001, the EPA adopted a new standard and public water systems must comply with the 10 ppb standard beginning January 23, 2006. See EPA, National Primary Drinking Water Regulations; Arsenic and Clarifications to Compliance and New Source Contaminants Monitoring, *Federal Register*, 66 (14), 6975, January 22, 2001.

8. Mossman, K.L. and Marchant, G.E., Radiation protection and the precautionary principle, *Risk: Health, Safety & Environment*, 13, 137, 2002.
9. Marchant, G.E. and Mossman, K.L., *Arbitrary and Capricious: The Precautionary Principle in the European Courts,* AEI Press, Washington, DC, 2004.
10. Weidemann, P.M. and Schütz, H., The precautionary principle and risk perception: Experimental studies in the EMF area, *Environmental Health Perspectives*, 113, 402, April 2005.
11. The DDT ban occurred prior to the formal establishment of the precautionary principle. Nevertheless, the ban on DDT is a clear example of extreme precaution. See Ray, D.L., *Trashing the Planet,* Regnery Gateway, Washington, 1990; Whelan, E.M., *Toxic Terror,* Prometheus Books, New York, 1993.
12. Marchant, G.E. and Mossman, K.L., Please be careful: The spread of Europe's precautionary principle could wreak havoc on economies, public health, and plain old common sense, *Legal Times,* XXVII (33), 58, August 15, 2005.

GLOSSARY

Absorbed dose The energy absorbed from exposure to ionizing radiation divided by the mass absorbing the radiation. Units: gray (Gy).

Acceptable risk (or dose) A risk or dose judged to have an inconsequential level of harm or injury and that is therefore considered to be safe.

Adaptive response Exposure of cells or animals to a low dose of radiation induces protective mechanisms against detrimental effects of a subsequent radiation exposure.

Alpha particle A particle consisting of two neutrons and two protons emitted in the process of radioactive decay from the nucleus of certain heavy radionuclides.

As low as reasonably achievable (ALARA) An approach to risk management that requires that risks should be made as low as reasonably achievable given economic and social constraints.

Assigned share The probability that a given cancer was caused by a previous exposure to a particular carcinogenic agent.

Becquerel (Bq) A unit of radioactivity equal to 1 disintegration per second.

Best available technology (BAT) An approach to risk management that requires that risks be reduced using the best available technology.

Bystander effect The capacity of cells affected directly by radiation to transfer biological responses to other cells not directly targeted by radiation. Responses in nontargeted cells can be beneficial or detrimental.

Cancer A disease characterized by uncontrolled growth of abnormal cells. Many cancers are malignant by having the capacity to spread to other parts of the body through the process of metastasis. There are more than 100 types of human cancer.

Carcinogen A biological, chemical, or physical agent that may cause cancer.

Collective dose A measure of population exposure obtained by summing exposures for all people in the exposed population. Units: person-sievert, person-rad.

Consistency principle Risk-reduction measures should be consistent with measures already adopted in similar circumstances.

Cosmic radiation A component of the natural background radiation consisting of high-energy ionizing radiation from outer space.

Cost-benefit analysis An approach to risk management requiring that risks, benefits, and costs be quantified so that they may be weighed against each other.

Countervailing risk Unintended risk that may occur as a consequence of management of target risks.

Deterministic effects Effects whose severity is a direct function of radiation dose. These effects are associated with a threshold dose that is effect dependent.

Dose proportion A dimensionless quantity calculated by dividing the measured or calculated dose by an appropriate reference dose, such as a regulatory dose limit or natural background level.

Effective dose The sum of the equivalent doses in all tissues and organs of the body weighted for sensitivity to radiation. Units: sievert (Sv).

Electromagnetic radiation Radiation propagated through space or matter by oscillating electric and magnetic waves. Electromagnetic radiation travels at the speed of light in a vacuum. Examples include gamma rays, x-rays, visible light, radiofrequency, and microwaves. Types of electromagnetic radiation are distinguishable by their frequency or wavelength.

Equivalent dose The absorbed dose averaged over a tissue or organ and weighted for the quality of the type of ionizing radiation.

Gamma rays High-energy electromagnetic radiation from the nucleus of some radionuclides.

Genomic instability The progeny of irradiated cells show an increased occurrence and accumulation of mutations, chromosomal aberrations, or other genetic damage for many senerations.

Gray (Gy) A unit of absorbed dose equal to 1 joule per kilogram.

Hertz (Hz) A unit of frequency equal to 1 cycle per second.

Hormesis A toxicologic phenomenon characterized by stimulation or beneficial effects at low doses and inhibition or detrimental effects at high doses. The dose response is nonmonotonic and is J-shaped or inverted U-shaped.

Hypothesis A hypothesis is a conjecture usually in the form of a question that is the basis for the design of experiments to test a theory.

Ionizing radiation Radiation of sufficient energy to dislodge electrons from atoms in the absorbing material. Ionizing radiation includes gamma rays, x-rays, neutrons, and alpha particles but excludes low-energy electromagnetic radiation such as radiofrequency radiation and visible light.

Latent period The period of time between exposure to a disease-causing agent and appearance of the disease. The latent period ranges from a few years to several decades for cancers caused by ionizing radiation.

Life span study (LSS) Epidemiologic study of the Japanese survivors of the atomic bombs. The sample includes approximately 85,000 persons for whom detailed radiation dose and medical information are available.

Linear no-threshold dose response A monotonic dose-response curve that is linear without a dose threshold. The slope of the dose-response curve is constant.

Meta-analysis The process of combining results from several epidemiologic studies that address related research hypotheses to overcome the problem of reduced statistical power in individual studies with small sample sizes. The results from a group of studies can allow more accurate data analysis.

Model A model is a physical or mental construct of a process or phenomenon. Models connect data to theory and facilitate theory development by conceptualizing underlying processes or structures.

Monotonic dose response A dose response characterized by increasing response with increasing dose.

Mutagen A chemical or physical agent that causes a permanent change in DNA (mutation) or causes an increase in the rate of mutational events.

Natural background radiation Radiation from natural sources including cosmic radiation from outer space, terrestrial radiation from naturally occurring radionuclides in the earth's crust, and naturally occurring radionuclides deposited in the body by ingestion or inhalation.

Neutron An uncharged subatomic particle in the atomic nucleus.

Nondiscrimination principle Risk reduction measures should not be discriminatory in their application, and thus comparable situations should not be treated differently.

Precautionary principle An approach to risk management that requires foregoing, postponing, or otherwise limiting a product or activity until uncertainty about potential risks has been resolved in favor of safety. The precautionary principle is often summarized by the phrase "better safe than sorry."

Proportionality principle Risk-reduction measures should be proportional to the seriousness of the potential risk.

Radioactivity The property of radionuclides to emit radiation spontaneously in the form of waves and particles. Radioactivity can be man-made or come from natural occurring radionuclides.

Radiogenic cancer Cancer caused by exposure to ionizing radiation.

Radionuclide A nuclide (a species of an atom characterized by specific numbers of neutrons and protons) that is unstable and emits radiation in the form of waves and particles.

Radon A radioactive noble gas that is the product of the radioactive decay of uranium in the earth's crust.

Relative risk The ratio of the incidence rate (or mortality rate) of disease in an exposed population to that in a control or reference population. A relative risk of 2 means that the incidence or mortality rate as a result of agent exposure is equal to the baseline rate.

Residual risk The risk that remains after risk-management actions have been implemented.

Risk The probability of a specific adverse outcome given a particular set of conditions.

Risk assessment A process to describe and estimate the probability of an adverse health outcome from exposure to an environmental agent and includes hazard identification, exposure assessment, dose-response assessment, and risk characterization.

Risk coefficient The increase in the incidence or mortality rate per unit dose.

Risk estimate The number of cases or deaths that are expected to occur in an exposed population per unit dose for a specified exposure scenario and expression period (often expressed as an annual estimate or lifetime estimate).

Risk management A process of analyzing, selecting, implementing, communicating, and evaluating strategies to reduce risks.

Risk offset A type of risk trade-off where the target and countervailing risks are the same and occur in the same population.

Risk substitution A type of risk trade-off where different target and countervailing risks occur in the same population.

Risk trade-off Efforts to manage the target risk may lead to increases in other countervailing risks.

Risk transfer A type of risk trade-off where the target and countervailing risks are the same but occur in different populations.

Risk transformation A type of risk trade-off where different target and countervailing risks occur in different populations.

Sievert (Sv) A unit of effective dose and equivalent dose. One Sv is equal to 1 joule per kilogram.

Single nucleotide polymorphism (SNP) Specific gene codon variations that result in a change of a single amino acid inserted at a specific site in the protein product of the gene.

Source term The amount and type of radioactive material released into the environment.

Stochastic effects Random events leading to effects whose probability of occurrence is a direct function of radiation dose. Stochastic effects are not believed to have a threshold dose.

Sublinear dose response A monotonic dose response curve that is concave upward without a dose threshold. The slope increases with increasing dose.

Supralinear dose response A monotonic dose response curve that is convex upward without a dose threshold. The slope decreases with increasing dose.

Target model Biological effects of radiation are thought to occur as a result of hits on cellular targets. Targets behave independently and once hit the target is incapable of repairing the damage. The target model is the underlying biophysical basis for LNT theory.

Target risk The particular risk that is the subject of risk management.

Theory A set of principles or statements to explain a collection of data, observations, or facts. Statements underlying a theory are often expressed in mathematical terms with biologically meaningful parameters.

Threshold dose A dose below which no response is observed.

Threshold dose response A dose response characterized by a threshold dose below which no response is observed.

X-rays High-energy electromagnetic radiation often produced by bombarding a heavy metal target with high-energy electrons in a vacuum.

Zero risk tolerance Any activity or product that involves any risk at all should be rejected.

Index

A

Absorbed dose, 137
Acceptable dose, 136, 139–140
Acceptable risk, 6, 7, 80
  approach, unacceptable costs and, 80
  problems addressed in defining, 102
Acoustic neuroma, cell phones and, 171
Acts of God, 90, 117
Adaptive response, 133
Agent(s)
  agent interactions, 131
  airborne, exposure to, 14
  exposure, evaluation of, 13
  IARC categories, 12
  waterborne, 14
AIDS, 86, 188
Airborne agents, exposure to, 14
Airlie House International Conference, 137
Airline accidents, 3
Air pollutants, 129
Alar, human cancer-causing effects of, 37, 38
ALARA (as low as reasonably achievable), 19, 80
  central point of, 39
  decisions, 95
  inappropriate applications of, 98
  philosophy
    precautionary approach vs., 183
    risk control using, 73
  process, acceptable doses and, 139
  program effectiveness, measurement of, 94
Alcohol
  consumption
    regulation of, 83, 111
    risks of, 81
  dual properties of, 33
  regulations, tightening of, 125
American Association of Physicists in Medicine, 102
American Cancer Society
  brain cancer estimates, 168–169
  estimates of national cancer burden, 122
  lung cancer death statistics, 153
American College of Radiology, 102
American Trucking Associations v. EPA, 119
Analog cell phones, 169
Anchoring bias, 115
Animal model, selection of, 16
Antidepressant drugs, 19
Antinuclear activists, LNT consequences and, 58–59
Apathy, catastrophe and, 90
Artificial sweeteners, 3, 18
AS, *see* Assigned share
As low as reasonably achievable, *see* ALARA
Assigned share (AS), 86
  calculations, 87
  radon, 86
  values, interpretation of, 87
Atkins® diet, 122–123
Atomic bomb survivors, 13
Automobile
  accident records, 4
  risks, 1
Avoiding risk, 129–147
  case against risk, 130–133
    agent–agent interactions, 131–132
    different risks, 130–131
    dose as surrogate for risk, 132–133
  case for dose, 133–136
  dose-based system of protection, 136–140
    acceptable dose, 139–140
    natural background, 137–139
    regulatory dose limit, 136–137
  management decisions based on dose proportion, 140–142
  notes, 145–147
  review of current system of radiation protection, 144–145
  simplification of radiation quantities and units, 142–144

B

Background radiation
  collective dose from, 55
  level uncertainties, 141
  limits of detection, 140
  natural, 14, 87, 112, 131, 137
Bands, 144
BAT, *see* Best available technology
BEIR Committees, *see* Biological Effects of Ionizing Radiation Committees
Below regulatory concern (BRC) policy, 102
Benzene, leukemia and, 86
Benzene Case, 118

Best available technology (BAT), 94, 95
Biological Effects of Ionizing Radiation (BEIR) Committees, 120
Birth, risk at, 1
Bone dose, 34, 35
Bovine spongiform encephalopathy (BSE), 172
Brain cancer, 169, 170
BRC policy, *see* Below regulatory concern policy
Breast cancer(s)
  genetic component of, 111
  incidence, LNT theory and, 48
British cell phone debacle, 174, 175
BSE, *see* Bovine spongiform encephalopathy
Bystander effect, 52, 132, 133

C

Caffeine, dual properties of, 33
Cancer, *see also specific types*
  burden
    HIV and, 110
    national, 122
  causing agents, 2
  chemotherapy and, 99
  damage, 52
  deaths, 111, 187
  development, current understanding of, 51
  dose-response curve for, 31
  environmental risk factors, 187–188
  genes, 71
  genetic susceptibility to, 85
  link between radiation and, 50
  mortality
    Japanese survivors of atomic bombings, 67
    risk assessment and, 114
  multistage process of, 51
  patients, screening of for radiation sensitivity, 86
  prevention, 122
  probability of induced, 12
  radiofrequency radiation and, 169
  radiosensitive groups and, 85
  rates, spontaneous, 71
  risk(s)
    animal tests to predict, 37
    coefficient, 71
    factors, classification of, 111
    PCBs and, 101
    ranking of, 110
    uncertainties of, 83
  spontaneous incidence of, 25
  treatment, ionizing radiation and, 100
Carcinogen(s), 2
  classification schemes, 12
  costs of regulatory control of, 123
  dose(s)
    measurement of, 75
    response, 15
  exposures, statistical probabilities for expression of, 129
  natural sources of, 138
  regulatory limits for, 74, 83
  risk assessment for, 47
Catastrophic natural events, 90
Catastrophic potential, 2, 3, 116
Causality, characteristics for establishment of, 11
Causation
  probability of, 86
  statistical association and, 9, 10
CEC, *see* Commission of the European Communities
Cell phone
  controversy, 172
  debate
    European, 177
    U.S., 177
  epidemiological studies, 169
  hazards, 178
  induced cancer risk, 171
  risks, social perceptions of, 177
  scare, U.K., 179
  useful purpose of, 179
Cell phone use, 167–181
  brain cancer and, 170
  concern about brain cancer, 168–172
  growth of, 167
  imprudent precaution, 173–177
  international calls, 177–179
  limiting, 175
  managing phantom risks, 172–173
  notes, 179–181
  risk of a risk, 173
Central nervous system (CNS) cancers, 169
CERCLA, *see* Comprehensive Environmental Response, Compensation and Liability Act
Cervical cancer, prevention of, 123
Chemical risk management, 94
Chemotherapy, cancer and, 99
Chernobyl nuclear power plant accident, 38, 55, 59
  dose proportion for, 185
  public views of, 114
Children
  cancer following diagnostic x-ray, 56
  cell phone use by, 173, 176
  CT studies, 150
  leukemia and, 69, 91
  Oxford Survey of Childhood Cancers, 69
  protection of, 89
  sensitivity of, 85
Chlorine Chemistry Council v. EPA, 39, 119

Cigarette(s)
causal evidence for, 11
consumption, regulation of, 111
packs, health warnings on, 2
regulations, tightening of, 125
smokers, worries of, 21
smoking
cancer mortality burden and, 110
lack of regulation of, 109
Clean Air Act, 95, 119
CNS cancers, *see* Central nervous system cancers
Cold War
development of nuclear weapons during, 25
nuclear weapons tests conducted during, 124
Collective dose, LNT theory and, 55
Colon cancer, 51
Columbine High School shootings, 179
Commercial airline travel, 20
Commission of the European Communities (CEC), 174
Communication
radon risks, 149, 150
risk, 19, 20
use of dose proportions in, 186
use of risk comparison in, 159
Comprehensive Environmental Response, Compensation and Liability Act (CERCLA), 83, 94
Computerized tomography (CT), 56, 83, 150, 186
Computer models, exposure assessment and, 14
Conceptual models, theory development and, 27
Confidence intervals, decision making and, 74
Conflicts of interest, perception of risk and, 92
Consistency, principle of, 174
Control over risk, 116
Cosmic radiation, 14, 139
Costs and benefits of action, 174
Cross-species extrapolation, 5, 11, 25, 40
CT, *see* Computerized tomography
Cyclamates, human cancer-causing effects of, 37, 38

D

Data
confidence in, 40
dynamic relation between theory and, 27
scientific, misguided interpretations of, 91
traffic fatality, 4
DDREF, *see* Dose and dose-rate effectiveness factor
DDT
discontinued use of, 80, 88
pesticide alternatives to, 7
precautionary principle and, 190
Delaney Clause, 6
Diet, cancer deaths and, 187
Dietary factors, cancer mortality burden and, 110
Dietary habits, 2
Diet control programs, 122
Digital cell phones, 169, 177
Dirty bomb, 185
DNA
damage, lung cancer and, 154
dependent cell damage, 54
model, Watson and Crick's, 28
radiation damage and, 49
DOE, *see* U.S. Department of Energy
Dose(s)
absorbed, 137
acceptable, 136, 139
based system, drawback to, 136
collective, 55
comparison, natural and man-made exposures, 191
direct measurement of, 133
estimates, risk estimates vs., 129–130
extrapolation
problem, magnitude of, 37
risk prediction and, 5
uncertainty and, 37
limit(s)
matrix, 137
occupational, 135
regulatory, 136
proportion(s), 183
concept, 184
environmental, 184
health risk communication using, 141
medical, 185
radiation protective actions triggered by, 140
public concept of, 134
reconstructions, 134
reduction, economic constraints, 81
risk coupling, 79, 80
standards-setting process, 134
threshold, absence of, 31
uncertainties, risk uncertainties vs., 134
whole-body equivalent, 130
Dose and dose-rate effectiveness factor (DDREF), 52
Dose-response
assessment, uncertainties of, 21
curve, 15
LNT theory and, 48
risk-communication strategies and, 25
risk management and, 39
studies, 16
sub-linear, 32

Driving
cell phone use while, 175, 177
drunk, 3
Drug(s)
addicts, 134
companies, clinical research sponsored by, 93
contraindicated in pregnancy, 101
Drunk driving, 3

E

EC Communication
ionizing radiation criteria, 98
precautionary principle and, 97
EC Treaty, Maastricht amendments to, 96
EEOICPA, *see* Energy Employees Occupational Illness Compensation Program Act of 2000
Effective dose, 15, 106
EIS, *see* Environmental Impact Statement
Electromagnetic field(s)
concerns about hazards from, 178
intensities, from cell phone towers, 167
Electromagnetic spectrum, 168
Electrosmog, 178
Energy Employees Occupational Illness Compensation Program Act of 2000 (EEOICPA), 86
Environmental contamination, decisions on cleanup of, 65
Environmental hazards, human exposures to, 8
Environmental Impact Statement (EIS), 58
Environmentalism, 178
Environmental problems, public perceptions of, 158
Environmental Protection Agency (EPA), 5
Ad Council TV spot, 90
agents considered hazardous, 12
arsenic standard for drinking water, 188
Clean Air Act, 95
cleanup standard, 25, 38
definition of acceptable risk, 80
dose framework, 75
estimates of lung cancer deaths, 149, 150
Federal Guidance Report No. 13, 72
Indoor Radon Abatement Act of 1988, 157
mismanagement of Superfund, 163
National Ambient Air Quality Standards, 119
perceptions of environmental problems, 158
preferred theory of, 30
principal health endpoint of concern to, 110
radon advisory issued by, 159
radon risks, 130
remediation of high radon homes, 157
risk coefficients, 72
risk system, 83
ruling on radiation standards for Yucca Mountain, 99
use of LNT theory by, 156
EPA, *see* Environmental Protection Agency
Epidemiology, probabilities and, 66
EU courts, precautionary principle and, 175, 176
Exposure
assessment, 14, 17
sources of, 15
Extremist groups, role in risk prioritization, 120

F

FACA, *see* Federal Advisory Committee Act
FDA, *see* U.S. Food and Drug Administration
Federal Advisory Committee Act (FACA), 120, 121
Federal regulatory agencies, use of LNT by, 17
FOIA, *see* Freedom of Information Act
Food additive, banning of, 18
Food Quality Protection Act of 1996, 6
Fossil fuels, global warming and, 126
Freedom of Information Act (FOIA), 120

G

Games of chance, 66
Gamma-emitting radionuclides, 143
Genetic diseases, 84
Genetic mutations, 49
Genomic instability, 133
Germ-line mutations, 71
Global warming, 126, 190
Glossary, 193–196
Government intervention, risk prioritization and, 118

H

Hazard
human evidence of, 11
identification
definition of, 9
exposure conditions and, 11
Hazardous air pollutants, EPA standards for, 80
Health detriment outcomes, calculation of, 41
Health effects, dose rate and, 134
Health Physics Society, 53, 102, 139–140
Healthy worker effect, 98
Heavy smoker, 134
HHS, *see* U.S. Department of Health and Human Services
High-risk events, 6
HIV, 86, 110
Home radon readings, 159, 160

Hormesis, 32, 33
biphasic nature of, 39
dose response, 34
LNT debate and, 42
risk-management objective under, 40
Human
carcinogens, costs of regulatory control of, 123
error, zero risk and, 89
papillomavirus vaccines, 123
radiation exposure, largest source of, 137
Hurricane Katrina, 90, 117
Hypothesis testing, 28

I

IARC, *see* International Agency for Research on Cancer
ICRP, *see* International Commission on Radiological Protection
Immediacy of risk, 116
Inappropriate theory, 26
Indoor Radon Abatement Act of 1988, 157, 163
Industrial Union Department AFL-CIO v. American Petroleum Institute, 118
Information agents, public trust in, 113
International Agency for Research on Cancer (IARC), 12
International agreements, precautionary principle, 96
International Atomic Energy Agency, 38
International Commission on Radiological Protection (ICRP), 28, 42, 50, 129
LNT controversy, 53
Main Commission, 144
preference for average tissue dose, 143
radiation protection policies, 162
risk-based methodology, 131
Ionizing radiation
cancer-causing effects of, 5
cancer treatment and, 100
carcinogenesis and, 54, 84
costs of reducing environmental health effects of, 124
criteria, EC Communication and, 98
cross-species extrapolation and, 36
exposure
health effects from, 40
individual's sensitivity to, 85
largest natural source of, 163
genomic instability and, 133
health risks, independent estimates of, 120
human experience and, 11
individual interaction with, 14
largest source of human exposure to, 149
leukemia and, 86
major source of, 14
mammography and, 111
measurement of, 14
probabilities of occurrence, 66

J

Japanese survivors of atomic bombings, 5, 31, 50, 66, 135

K

Knowledge of risk, 116

L

Leukemia, 86
children dying of, 91
dose response, 52
incidence, 31, 32
sublinear dose responses, 132
Lewis, E.B., 50
Life insurance premiums, 111
Life Span Study (LSS), 66, 67
Limerick Nuclear Power Plant, 150
Linear no-threshold (LNT) theory, 47–64
alternative predictive theories, 53
attraction of, 48
cancer risk assessment and, 132
collective dose and, 55
consequences, 58–60
controversy, 53–55
elements of debate, 54
question of thresholds, 54
repair of radiation damage and cellular autonomy, 54–55
debate, major issues driving, 54
decision to use, 53
dose response, 16
curve, 48
indoor radon studies and, 154
miner studies and, 155
equation of, 27
federal regulatory agency use of, 17
flawed concept of, 145
forced scientific justification of, 53
interpretation of, 95
notes, 60–64
nuclear community's support of, 54
Ockham's razor and, 70
radiogenic cancer risk and, 187
radon risks and, 149
risk per unit dose, 28
single order kinetics of, 50–51
straw that broke the camel's back, 51

theory of choice, 48–52
uses and misuses, 55–58
childhood cancer following diagnostic x-ray, 56–58
estimation of health effects of fallout from Chernobyl reactor accident, 56
public health impacts from radiation in modern pit facility, 58
U.S. Nuclear Regulatory Commission use of, 30
LLRWPA, *see* Low-Level Radioactive Waste Policy Act
LNT theory, *see* Linear no-threshold theory
Low-dose risks, EC communication and, 98
Low-Level Radioactive Waste Policy Act (LLRWPA), 59
Low risk
definition of, 5
events, examples of, 5–6
LSS, *see* Life Span Study
Lung cancer
burden, strategy for reducing, 164
deaths, 11
American Cancer Society statistics on, 153
EPA estimates of, 149, 150
smoking and, 153
DNA damage and, 154
mortality
radon and, 86
rate of, 130
probability of dying from, 114
radon exposure and, 142
radon gas and, 131
uranium mining and, 151–152

M

Mad cow disease, 172, 178, 179
Malaria, terminated use of DDT and, 80, 88–89
Mammography
ionizing radiation and, 111
LNT theory and, 48
Manhattan project, 135
Maximum contaminant level goal (MCLG), 119
MCLG, *see* Maximum contaminant level goal
Medical radiation, 13, 14
Medical therapy decisions, conflicts of interest in, 93
Medications, risks of side effects, 19
Misinterpreted risks, 4
Misplaced priorities, 109–128
factors in prioritization, 112–121
court actions, 118–119
influence of stakeholder groups, 119–121
management capacity, 117–118
public perception of risks, 114–117
scientific evidence, 113–114
monetary costs, 123–125
characterization of waste destined for WIPP, 125
environmental cleanup at Nevada test site, 124
notes, 126–128
priorities and realities, 110–112
real risks and reordering priorities, 121–123
risks in perspective, 125–126
Molecular photocopying, 43
Motorola, Inc., 171
Muller, Hermann J., 48–49

N

National Academies
advice provided by, 121
BEIR VII Committee, 42
National Council on Radiation Protection and Measurements (NCRP), 28, 42, 47, 53, 131
National Highway Traffic Safety Administration, 20
National Nuclear Security Administration (NNSA), 58
National Radiological Protection Board (NRPB), 173
National Research Council, 8
BEIR Committee, 31, 120, 135, 156, 162
definition of risk management, 17
Red Book, 8, 23, 18
risk-assessment process, 9
National Toxicology Program, HHS, 13
National wealth, risk-management strategies and, 92
Natural background radiation, 14, 87, 131, 137
cancers attributable to, 112
collective dose from, 55
Nazi national health program, 92
NCRP, *see* National Council on Radiation Protection and Measurements
Nevada Test Site (NTS), 124
Newman v. Motorola, Inc., 171
New Yorker articles, 91
Nicotine, dual properties of, 33
NNSA, *see* National Nuclear Security Administration
Nondiscrimination, principle of, 174
Nonmelanoma skin cancers, 4
Nontargeted delayed effects, 133
No safe dose concept, 50
NRPB, *see* National Radiological Protection Board
NTS, *see* Nevada Test Site

Nuclear power
electricity supply and, 126
plant(s)
operations, radiation exposure from, 15
radionuclide emissions from, 9
Nuclear weapons stockpile, U.S., 58

O

Obesity, 81, 187
Occupational dose limits, 135
Occupational Safety and Health Administration (OSHA), 95, 119
Ockham's razor, LNT theory and, 70
Oncologists, objectives in curative therapy, 100
Optimization principle, radiation protection, 80
OSCC, *see* Oxford Survey of Childhood Cancers
OSHA, *see* Occupational Safety and Health Administration
Over-the-counter medication, safety of, 6
Oxford Survey of Childhood Cancers (OSCC), 69

P

PAG, *see* Protective action guide
PCBs, *see* Polychlorinated biphenyls
PCR, *see* Polymerase chain reaction
Pediatric CT examinations, 57, 150, 186
Perceived risk, 81
Personal behaviors, risks of, 81
Personal exposure monitoring, 14
Personal judgments, risk management, 93
Pesticides, 6, 7
cancer burden and, 110
regulatory analysis, 101
Phantom risks, 172
Physical inactivity, deaths attributed to, 121, 187
Pluralism, 82
Policy makers, risk management and, 91
Pollutants, cancer burden and, 110
Polluters, blaming, 90
Polychlorinated biphenyls (PCBs), 100, 138
Polymerase chain reaction (PCR), 43
Popper, Karl, 28
Population
health detriment, 41
risks, 57
PR campaign, 183–192
notes, 191–192
precaution, 189–191
prioritization, 187–189
proportion, 184–187
Precautionary principle, 95
ambiguity of, 99
application of, 98
cause-and-effect relationships and, 174
cell phone use by children, 173
DDT and, 190
EU courts and, 175
extreme application of, 176
implementation of, 96, 189
inconsistencies in, 97
international agreements, 96
phrase summarizing, 96
provisional, 174–175
Pregnancy, drugs contraindicated in, 101
Pregnant women, 85
morning sickness in, 101
protection of, 89
Presidential/Congressional Commission on Risk Assessment and Risk Management, 18
Priorities, *see* Misplaced priorities
Prioritization
process, influence of science in, 109
public health protection and, 183
Probability of causation, 86
Proportionality
precautionary principle and, 174
test, 175
Prostate cancer, 187–188
Protective action guide (PAG), 71
Public
apathy to radon, 163
health protection, primary goal of, 187
information, distortion of risk and, 90
interest groups, legal actions by, 92
involvement in risk management, 91
participation in decision-making process, 120
Public policy
imprudent, 91
risk-management agenda and, 109

Q

Quantitative risk estimation, 40

R

Radiation, *see also* Ionizing radiation
action, target model of, 49
background, 14
cosmic, 14, 139
damage, DNA and, 49
exposure, sources of, 15, 150, 186
fear of, 59, 60, 79
genetic effects of, 50
hormesis proponents, 40
largest anthropogenic source of, 60
medical, 14

protection
decision making, dose proportions in, 140
optimization principle, 80
practice, ALARA and, 19
radiofrequency, 168, 169
technology-derived, 139
Radioactive waste disposal, 1, 59
Radiofrequency (RF), 167, 169
Radiological dispersal device, 185
Radionuclide(s)
composition, of Earth's crust, 112
gamma-emitting, 143
risk coefficients for, 72
Radiosensitive groups, 85
Radon
billboard messages, 159
distribution of in homes, 152
effective dose for, 15
gas, 14, 15, 131
guidance, most restrictive, 163
home remediation costs, 161
largest source of data on health effects of, 155
lung cancer
mortality, 86
risks, 155
measurements, variability in, 153
problem, 156
public apathy to, 163
regulation, 162
remedial action program, 151
risks, communication of, 149, 150
tests, apathy and, 90
Radon exposure, 149–166
health hazards of radon, 153–156
human exposure to radon, 152–153
lung cancer and, 142
notes, 164–166
question of public health hazard, 156–163
economic impacts, 160–162
national/regional differences, 162–163
perceptions and fears, 158–160
public health, 156–158
smoking, 132, 163–164
Watras case, 150–152
Rationality, risk-management decisions and, 82
Reading Prong, high soil levels of radon in, 153
Red Book, National Research Council, 8, 18, 23
Regulatory dose limit, 136
RF, *see* Radiofrequency
Right dose-response theory, 26
Risk
acceptable, 6, 7, 21, 80, 102
analysis, value judgments and, 82
assessment
elements of, 8
federal agencies performing, 8
questions of, 7
uncertainty and, 8
assigned share of, 86
avoidance behavior, 59
cancer, 83
catastrophic potential of, 116
characterization, 16, 17
coefficient, 28, 52
communication, 19, 20, 40
comparison, communication to public using, 159
control over, 116
definition of, 3
distortion of, 90
dose as surrogate for, 130, 132
epidemiological-based, 70
estimation, quantitative, 40
expression of, 4, 65
immediacy of, 116
insignificant, 186
knowledge of, 66, 116
large, 83
low, definition of, 5
magnitude of, 3
misconceptions, 129
misinterpreted, 4
offset scenario, 100
perception, 2, 81, 114
differences in, 20
factors impacting, 115
as obstacle to technological progress, 60
phantom, 172
prevention strategies, public health benefit of, 1
prioritization, 21, 112
probability and consequence, 3
public perception of, 114
public understanding of, 2
quantification, 184
reduction, economic costs of, 161
risk trade-off, 6
scientific expression of, 20
small, 83, 91
society's prioritization of, 82
substitution, 100
technological, 2
countervailing risks and, 109–110
perception of, 59
public views of, 115
trade-off problem, 99
transfer, 101
transformation, 101
trivial, 38
unavoidable, 1

very small, interpretation of, 71
voluntariness of, 116
zero, 40, 88, 89
Risk management, 79–107
approaches to, 80
challenges, 101–103
decision(s), 82
framework for making, 18
social and political factors, 82
definition of, 17
dose-response curve and, 39
goal, 36, 79, 109
management strategies, 94–99
as low as reasonably achievable, 94–95
best available technology, 95
precautionary principle, 95–99
management triggers, 81–94
social triggers, 87–94
technical triggers, 82–87
notes, 103–107
precautionary principle, 96
public involvement in, 91
risk–risk trade-offs and unintended consequences, 99–101
risk offset, 100
risk substitution, 100–101
risk transfer and risk transformation, 101
socioeconomic implications, 18
stakeholder involvement and, 18
trigger, prime, 88
Risky business, 1–24
consequences of risk, 3–6
damage control, 17–19
danger, 9–13
exposure, 13–15
health risk, 15–16
notes, 21–24
perception is reality, 19–21
risks, 16–17
safety without risk, 6–7
what's risky, 7–9
Rocky Flats Plant, 58

S

Saccharin, 7, 12, 16, 22
hormesis and, 34
human cancer-causing effects of, 37, 38
risk assessment of, 113
Safety
definition of, 6, 88
determining, 7
selling of, 25
Safety point, threshold dose establishing, 40
Scientific data, misguided interpretations of, 91
Scientific guesswork, 25–45
limitations and uncertainties, 36–38
making the right choice, 26–29
notes, 42–45
predictive theories in risk assessment, 30–36
hormesis, 32–33
linear no-threshold theory, 30–31
sublinear nonthreshold, 31
supralinear, 32
threshold, 34–36
quantifying risk at small doses, 41–42
risk management and risk communication, 39–41
speculation versus reality, 38
Scientific inquiry
hypothesis testing and, 28
subjective process of, 29
Scientific risk assessment, public view of, 113
Seat belt restraints, 2
Secular trend analysis, 170
SEER cancer database, *see* Surveillance, Epidemiology and End Results cancer database
September 11 (2001), 100, 179
Sierra Club, 178
Signal-to-noise ratio, 67, 69
Single nucleotide polymorphisms (SNPs), 84
Small risks, media focus on, 91
Smog, 129
Smoke-free workplace, 125
Smoking
deaths attributed to, 121, 187
health effects of, 130
lung cancer deaths and, 153
risks of, 1, 81
SNPs, *see* Single nucleotide polymorphisms
Social regulations, risk prioritization and, 118
Social triggers, values underlying, 88
Socioculturalism, 82
Soils, uranium concentrations in, 153
South Beach diet™, 122–123
Southwest Compact, 59
Special interest groups
agendas of, 120
LNT consequences and, 58–59, 60
Stack effect, 164
Stakeholder
groups, risk-management strategies and, 91, 92
perspective(s)
risk-management agenda and, 109
risk-management decisions and, 119
Standards-setting organizations, 33, 92
Statistical association, causation and, 9, 10
Stereotyping, 115
Stewart inquiry, 172

Sublinear theory, 31
Sumatra tsunami (2004), 118
Superfund, *see also* Comprehensive Environmental Response, Compensation and Liability Act
  EPA mismanagement of, 163
  sites, remediation of, 8
Supralinearity, 32, 36
Surveillance, Epidemiology and End Results (SEER) cancer database, 170
System of protection, goal of, 136

T

Tamper-resistant packaging, 7
Target model of radiation action, 49
Technological risks, 2
  countervailing risks and, 109–110
  perception of, 59
  public views of, 115
Technology-derived radiation, 139
Theory(ies), *see also* Linear no-threshold theory
  accuracy of predictions, 29
  classification of, 30
  confidence in, 29
  definition of, 28
  dynamic relation between data and, 27
  inappropriate, 26
  selection, 26, 47
  sublinear, 31
  test of, 28, 29
  threshold, 34
Third World countries
  infectious diseases in, 92
  worries in, 116
Three Mile Island (TMI) accident, 59, 69, 89, 114
Threshold theory, 26, 34
Thyroid
  cancer, probability of dying from, 114
  gland, iodine uptake by, 131
Timofeeff-Ressovsky, Nikolai, 49
Tissue-weighting factors, 131, 142
TMI accident, *see* Three Mile Island accident
Tobacco
  deaths attributed to, 121, 187
  tax revenue, lost, 125
  use, 4
Traffic fatality data, 4
Transuranic waste-management program, 125
Trivial risks, 38
Tsunami, most devastating in recorded history, 118
Tylenol® scare, 7, 115

U

Uncertain risk, 65–77
  another approach, 74–75
  notes, 75–77
  range of doses, 66–70
  risk assessment considering uncertainty, 70–72
  uncertain choices, 72–74
Uncertainty
  cross-species extrapolation and, 11, 17, 36
  distinctions between knowledge and, 65
  dose extrapolation and, 37
  magnitude of in risk estimates, 26
  risk assessment and, 8
United Nations
  Rio Declaration on Environment and Development, 96
  Scientific Committee on the Effects of Atomic Radiation (UNSCEAR), 162
United States
  government, regulatory program costs of, 123
  major causes of death in, 121, 122
  nuclear weapons stockpile, 58
UNSCEAR, *see* United Nations Scientific Committee on the Effects of Atomic Radiation
Uranium miners, lung cancer and, 151–152
U.S. Department of Energy (DOE), 86, 92
  argument over Yucca Mountain radioactive waste repository, 93
  defense-related transuranic waste cleanup by, 125
  guidelines to remediate high radon homes, 157
  National Nuclear Security Administration, 58
  Savannah River Site, 101
U.S. Department of Health and Human Services (HHS), 13
  National Toxicology Program, 13
  Report on Carcinogens, 13
U.S. Food and Drug Administration (FDA), 101
U.S. NRC, *see* U.S. Nuclear Regulatory Commission
U.S. Nuclear Regulatory Commission (U.S. NRC), 19, 102, 110, 131
  BRC policy, 102
  cleanup standards, 38
  framework for decision making, 19
  preferred theory of, 30
  principal health endpoint of concern to, 110

V

Value judgments, 82
Volcanic eruptions, 117
Voluntariness of risk, 116

W

Waste Isolation Pilot Plant (WIPP), 125
Water, dual properties of, 33
Waterborne agents, exposure to, 14
Watras case, 150, 160
Weight-reduction programs, 122
Whole-body equivalent doses, 130
Wingspread Statement, 96
WIPP, *see* Waste Isolation Pilot Plant
World Health Organization, 12, 88
World Trade Center, 100

X

x-ray(s)
  fear of, 99
  mutagenic properties of, 49
  procedures
    medical, 13
    refusal of diagnostic, 60

Y

Yucca Mountain radioactive waste repository, 93, 99

Z

Zero probability problem, 66
Zero risk, 40
  approaches to achieving, 88
  human error and, 89
  tolerance, 6, 183, 189